Biosemiotics

Volume 19

Aims and Scope of the Series

Combining research approaches from biology, semiotics, philosophy and linguistics, the field of biosemiotics studies semiotic processes as they occur in and among living systems. This has important implications and applications for issues ranging from natural selection to animal behaviour and human psychology, leaving biosemiotics at the cutting edge of the research on the fundamentals of life.

The Springer book series *Biosemiotics* draws together contributions from leading scholars in international biosemiotics, producing an unparalleled series that will appeal to all those interested in the origins and evolution of life, including molecular and evolutionary biologists, ecologists, anthropologists, psychologists, philosophers and historians of science, linguists, semioticians and researchers in artificial life, information theory and communication technology.

More information about this series at http://www.springer.com/series/7710

Pauline Delahaye

A Semiotic Methodology for Animal Studies

Springer

Pauline Delahaye
Paris-Sorbonne University
Paris, France

ISSN 1875-4651 ISSN 1875-466X (electronic)
Biosemiotics
ISBN 978-3-030-28812-9 ISBN 978-3-030-28813-6 (eBook)
https://doi.org/10.1007/978-3-030-28813-6

This Springer imprint is published by the registered company Springer Nature Switzerland AG.
The registered company address is: Gewerbestrasse 11, 6330 Cham, Switzerland

Acknowledgements

While it looks like a very solitary work, a scientific book, especially when closely linked to a thesis, as it is the case here, is never the result of an isolated mind. Many people have brought me their advice, help or support when I needed it most. The list is not exhaustive, and I most likely have forgotten many, but I insist on thanking them here nonetheless.

- Kalevi Kull and Paul Cobley, who made me the proposition of writing this book, submitted the project to the editor and completely trusted me for the rest of this endeavour.
- The Springer Nature team, especially Karthika Menon, who accompanied me throughout the writing of this book so that it would take the shape that the reader is currently holding.
- Astrid Guillaume from the Université Paris-Sorbonne, who directed my thesis, for her attention to details I never noticed and for her theories in semiotics on intertheoricity that allowed me to make my model a reality by working on theories from different disciplines.
- Very special thanks to Pierre Ghislain, my translator and old university friend, who accomplished a tremendous amount of work on the translation and correction of my words for this book, which would simply not exist without his work.
- Georges Chapouthier, research director at the CNRS and president of the jury, member of my jury at my midterm report, who provided many lines of approach and reflection, without whom, the ethological reach of this semiotic methodology would not be the same.
- Professor Thomas Broden, from Purdue University, member of my jury at my thesis defence and my midterm report, for opening new, unsuspected bibliographical avenues and providing references that were unobtainable in France.
- Claude Béata, veterinarian and doctor in behaviour studies and member of the jury whose books provided me with a large amount of the ethical thinking part of this book and my thesis.
- Other members of the jury, Ekaterina Velmezova and Anne-Claire Gagnon, for their comments that helped me refine this methodology.

- Professor Jean-Baptiste Rauzy, director of the *Ecole Doctorale Concepts et Langage*, who facilitated my registration and reception, and allowed me, through the *Ecole Doctorale*, to take part in colloquiums far from France, leading me to meet other biosemiotics and zoosemiotics specialists spread around the world who I never would have met otherwise.
- Professor Jacques Dürrenmatt, director of *EA 4509*, and Claude Montacié, director of UFR de Sociologie et Informatique pour les Sciences Humaines for the Université Paris Sorbonne, without whom, the works at the origin of this book would have never been successfully completed.
- My colleagues at Sorbonne Université, for their fruitful discussions that nourished my work and my thinking.
- My colleagues in computer science, who forced me to make my thinking accessible to non-specialists and greatly influenced my writing style.
- My students in the French Language and Information Technology curriculum, for their interest and their questions that led me to more research as well as making my writing more educational and clearer.
- My intern at the *Société Française de Zoosémiotique*, Gabrielle Zabus, who helped with translating and checking the abstracts at the beginning of each chapter.
- A thought in memory of Vinaigrette, who led me towards the study of non-human creatures, without whom, I clearly would not be doing what I am doing today.

Contents

About the Author

Pauline Delahaye is a doctor of linguistics from the University Paris-Sorbonne and is a specialist in zoosemiotics. She received her PhD in 2017. Her thesis (*Des signes pour le dire: étude sémiotique des émotions complexes animales*, Presses Universitaire de Rennes, 2019) is the first on the subject to be defended in France.

She is a member of the Board of Directors for the Société Française de Zoosémiotique and for the Grand Paris Sémiotique, and as such, she actively strives to develop semiotic studies and helps with the establishing of biosemiotics and zoosemiotics in France.

After teaching at the Sorbonne Université for 2 years, she is now involved in the scientific mediation with the general public, for educational associations (*Prométhée Education*), cultural associations (*Pariscience, Science and Télévision*) as well as for institutes and foundations of general interest (*Institut Jane Goodall*).

List of Figures

List of Tables

Chapter 1
Introduction and Purpose

Abstract This chapter presents the rest of the book and serves as its introduction.

It starts by describing the creation process of this methodological work, and explains this choice of a semiotic study of the animal, as well as the type of corpus used and the expected societal and academic impact.

Then it discusses the reason behind introducing semiotics to the field of animal studies, relating the history of semiotics to the history of animal studies.

Lastly, it briefly addresses some methodological questions on the difficulties that come with studying a subject from different academic fields, adding new documentation categories into corpora or taking into account the various biases and ideologies that may appear.

The chapter ends by laying out the outline of the book.

Keywords Animal studies · Semiotics · Academic fields · Semiotic model · Corpus

This methodological model project occurs in a particular intellectual and societal context, one of reflecting upon the "animal person". The question of what an animal is, and what its cognitive, social and emotional capabilities are, is now more under the spotlight than ever; and it doubles up as a highly complex intellectual and philosophical question.

In this context, it seemed to me that the challenge of this methodology used to solve these questions was more important than ever. This is particularly true for certain aspects of animal life that remain mostly unexplored, such as their emotional and social lives, or the impact that long-term memory and the accumulation of recollections have on them. All of these fields entail an Umwelt methodology that is still rare in ethology. I also saw the subject of methodology as important because the notions of mental world and subjectivity allowed me to question a position that I saw as problematic. It was the study of the animal as a comparative element to humans, instead of studying, on one hand, the animals – including humans – themselves and by themselves, and on the other hand, the general continuity of life. I sincerely believe that shifting from the first position to the second would allow more relevant and less tense studies on many fascinating yet controversial subjects that

P. Delahaye, *A Semiotic Methodology for Animal Studies*, Biosemiotics 19,
https://doi.org/10.1007/978-3-030-28813-6_1

are currently pushed aside, not due to lack of techniques or methods but rather due to ideological positions and a stranglehold from their discipline.

I therefore hope to bring three helpful contributions to three distinct fields.

The first relates to linguistics itself, as it has long considered the question of animal language a "non-question". The origin of this realisation of mine lies in reading the famous article by Benveniste (1966) on the bee dances. He begins his work by explaining that he examined this unique form of communication from that species because it had been proven that no sound of any kind in the animal kingdom conveyed any meaning. The resulting lack of interest in the animal question in the following years created some very damaging consequences.

The second contribution I hope to bring is to the field of life sciences, which, after many long conversations with ethologists, seems to have shown too much rigidity in its study methods on certain topics.

Finally, regarding the societal examination of the self and the animal "person", its place in the world and how it should be considered, as I discussed at the beginning of this general introduction, I hope this work will help move the debate forward and give the people concerned a theoretical tool to rely on so as to allow more enlightened stances. Of course, I do not claim that I will revolutionise all the knowledge on this topic, and there have been many new studies on animal intelligence in recent years. But I do believe this question has not been sufficiently explored from a semiotic point of view, even though it is at least just as important as the physiological aspect, especially for certain professions dealing with animal well-being or the preservation of species.

1.1 Creation Process

The current state of animal semiotics is complex – few studies have been published on the topic, and they need to be looked for in works in biology, primatology, cetology, etc. Another approach would be to take stock of the current state of both linguistics and ethology in all the fields capable of coexisting within both disciplines. This did not seem very relevant as it is not very useful for my approach.

However, if there have been actual founders of semiotics applied to ethology, such as Sebeok, it is clear that, nowadays, it is hard to find fully-fledged trends in this field, and harder even with a leader at their head.

The field itself has suffered from sometimes virulent criticism, arguing that it was far too complicated to put together a true semiotics of the animal kingdom, and that subjects that are hard to measure or codify such as language, thought or emotions could not be really important, or should not be taken into account out of concern for objectivity. On this last point, I fall in with the criticism that Yankelovich, under the pen name of Adam Smith (1972), makes of that state of mind in scientific research:

> The first step is to measure whatever can easily be measured. This is okay as far as it goes. The second step is to disregard that which can't be measured or give it an arbitrary quantitative value. This is artificial and misleading. The third step is to presume that what can't be measured easily isn't very important. This is blindness. The fourth step is to say that what can't be easily measured really doesn't exist. This is suicide.

As a result, the foundations of my research endeavour needed to be different from the academic models, while keeping great scientific rigour, so as to be coherent with an interdisciplinary thesis and pertinent for the different fields that will be called upon in this study.

With this in mind, I chose a double theoretical framework that will be attach itself to a corpus convincing to each disciplinary branch. To do so, I have created a semiotic theoretical framework from the works of theorists that seemed the most coherent with the concerns of life science researchers. This framework allows proceeding between different fields and different species.

1.1.1 About the Subject

If there have been few works in semiotics applied to animals (and therefore few methodological works), it is first and foremost due to academic tradition. Semiotics, at least in modern universities, is seen as a branch of linguistics. However, the main focal point of linguistics is language, which was long considered to be exclusive to man. The debate around animal "language" is nothing new. Benveniste (1966, p. 56) had already started (and closed it almost simultaneously) in tome I of *Problèmes de Linguistique Générale*:

> Appliquée au monde animal, la notion de langage n'a cours que par un abus de termes. On sait qu'il a été impossible jusqu'ici d'établir que des animaux disposent, même sous une forme rudimentaire, d'un mode d'expression qui ait les caractères et les fonctions du langage humain. Toutes les observations sérieuses pratiquées sur les communautés animales, toutes les tentatives mises en œuvre au moyen de techniques variées pour provoquer ou contrôler une forme quelconque de langage assimilable à celui des hommes ont échoué.
>
> Translation:
>
> Applied to the animal world, the notion of language exists only through an abuse of terms. We know that it has so far been impossible to establish that animals have even a rudimentary form of communication with the characteristics and functions of the human language. All serious observations done on animal communities, all attempts put together with various techniques to provoke or control a form of language that could be assimilated to man's have failed.

That was in 1966. That should prompt us to do three things. First off, to be cautious: the works of the Gardners could already cause one to second-guess that statement, and barely 10 years after that paragraph, the works of Irène Pepperberg (1999) on Alex, the grey parrot, would render it obsolete. They would only be the firsts of a long list, and little by little, the issue of using the term "language" would return, whether it was for primates or a more distant branch like cetaceans.

Secondly, it encourages us to make an important distinction between what we know not to exist (for example, articulated language in other primates) and what we do not know. Benveniste seemed to say that everything had already been tried in order to try and find a language worthy of its name in all observable animal species. However, if we take the example of the dog, an animal that has been at man's side for possibly millennia, the first study attempting to find a semantic aspect in a dog's barks only came out in 2002 (Yin 2002). It is a basis of rhetoric; it is scientifically impossible to prove the non-existence of something, we can only strongly suspect its absence when the experiments made to prove its existence fail and it is logical and wise to come round to the opinion of "strongly possible non-existence". In this case, Benveniste should have refrained from making such categorical claims.

Lastly, it also invites us to be vigilant towards the ideological stances of the authors of scientific literature and the constantly question them. In 1966, the works of the Gardners were already under way (Fouts would join them in 1967 to assist them) and it had been over 5 years since Goodall's reports were describing chimpanzees building and using tools in the wild, and transmitting these techniques to their young. Elements that could lead linguists towards a more moderate stance already existed. Why they were not taken into account would be an unnecessarily controversial issue; however, we do need to keep in mind that they were not taken into account and that, as a result, we should be cautious whenever we encounter any deep-seated stance in scientific literature.

While one could think that the advances of the last 50 years in biology, primatology and other fields would have made it clear that language is most likely not exclusive to humans and that we should rekindle that debate in language sciences, it evidently is not the case. This disciplinary bias has been previously discussed by Lieberman (1977, p. 5) when he was collaborating with Sebeok:

> The supposed uniqueness of human language seems to me to be an echo of the traditional Cartesian view. Many ethological, behavioral, and linguistic studies of the communications of animals that are otherwise faultless are limited by their unquestioning acceptance of this Cartesian premise.

However, researchers do not all always agree on the characteristics of what makes or does not make a language. This definition is part of those I deem particularly problematic, as they seem to create a disciplinary barrier that nobody appears to be able to define clearly. As a logical result, it will be one of the main talking points of my second and third chapters. It is also the reason why my subject is not animal language, and my method is not a linguistic method but rather a semiotic one. What I am proposing here is to study various forms of complex communicational manifestations in the animal kingdom, while giving the appropriate tools and methods to study them.

These manifestations include those that could be categorised as language, just as they include those linked to learning, skills and transmission, those linked to intelligence, conscience and cognition or even those linked to emotional life and experience. In order to find them a strong enough semiotic value, these manifestations must fall into at least one of the three categories. They must convey a specific message, or be the result of an intent to communicate, or carry complex content.

Those semiotic categories constitute what I have isolated in my thesis work, and called the notion of "saying" as opposed to the "language" category that I deemed too reductive and restrictive. I will of course have a deeper discussion about those categories when I will approach the issue of the definition of language in semiotics applied to non-human animals, but for the sake of clarity, I offer a brief description below.

As far as the specific message is concerned, there is obviously still the possibility of a misinterpretation, including in human languages; the message may not be specific enough so as to remove any ambiguity. That is not the issue. The issue is in the act of signifying itself; in producing a semiotic act that has a specific meaning that we may reasonably think will be interpreted as such. This view of the act of "saying" is not new; Eco had already discussed it in 1988:

> La seule chose qui paraît indiscutable, c'est l'activité de signification. Il est commun aux humains (et la zoosémiotique s'interroge pour savoir si ce n'est pas aussi le fait de nombreuses espèces animales) de produire des évènements physiques (ou d'avoir la capacité de produire des classes d'évènements physiques) qui viennent se substituer à d'autres évènements ou entités, physiques ou non, que les humains ne sont pas en mesure de produire dans l'acte de signification […]
>
> Les processus de signification seraient l'artifice indéfinissable que les êtres humains, dans leur impossibilité d'avoir le monde entier (réel et possible) à portée de main, mettraient en œuvre pour suppléer à l'absence des choses. (Eco 1988)
>
> Translation:
>
> The only thing that seems undisputable is the act of signifying. It is common, for humans (and zoosemiotics is wondering if that is not also the case in many animal species) to produce physical events (or to have the capacity to create classes of physical events) that substitute other events or entities, physical or not, that humans are unable to produce with the act of signifying.
>
> The processes of signifying would be the indefinable trick that human beings, facing the impossibility to have the entire (real and possible) world in their grasp, have set up to compensate for the absence of certain things.

Even under this simple aspect, the existence of the specific message reveals the importance of the notion of "saying", of the need to communicate something so important that it must be conveyed in a way that leaves as little room for misinterpretation as possible. The existence of a specific message in a species is an exterior sign of its cognitive complexity and the importance it gives to a good communication between its members.

For the aspect of voluntary communication, it is an important point to understand the effort that one individual can exert in their language performance. With this aim in mind, I studied the case of Koshik (Stoeger et al. 2012), a male elephant in a Korean zoo, capable of articulating a few Korean words (see Fig. 1.1).

This articulation does not come with a comprehension of the words he uses, however, as they do not seem to be used properly or in the right context. The part that truly interests us is the phonological aspect of his performance, which is particularly impressive. To understand why, it is important to know that the upper lip of

Fig. 1.1 Koshik the male elephant, and his carer in 2012

elephants cannot move like ours can, it is bound to the nasal bone and therefore does not allow for an articulated pronunciation, which Koshik understood. To compensate for this anatomical feature that was preventing him from producing the sounds he wanted, he used a complex compensation technique: he placed his trunk in his mouth and voluntarily mobilised it against his lower lip so as to use it as a moving upper lip. The technique was advanced enough to allow the elephant to articulate a few understandable words that could be frequently heard. Stoeger's team that studied his case is fairly certain that, while Koshik is not necessarily ignorant of the meaning of his words, he, at best, has a very imperfect use of them. However, the researchers underlined how exceptional the technique is, and confirm that the words were very much audible and recognisable.

This performance is impressive and can only make us wonder what motivates an individual to try and do something this complex. Some contextual elements point to an answer. Koshik had been alone since his arrival at the zoo, with no other members of his species around him. Conscious of the fact that his species needs a lot of social contact, his carer spent a lot of time with Koshik, but that did not seem to be enough. The performance of Koshik has a very specific, important goal: engaging in communication with this individual from another species by trying to use their communication channel in order to reinforce social interactions with them, request more of them, and have them understand that he needs them, even if the form of the message is imprecise and incorrect.

This aspect had already been made explicit by Eco in 1987; the act of communicating exists because there is an urge, a will, a search for collaboration from the emitter to the receiver. The idea is not new, but it had not yet been explored as a study approach for zoosemiotics. In its simplest form, this aspect is similar to the phatic function of language, is characterised by the means (cognitive, temporal, energy consumption…) put into motion by the sender to try and establish a communication, and/or by the receiver to try and perceive the message. Depending on the context, this aspect can tell us a lot about the motivations of the individual, that is to say, why they are willing to put in so much effort, and the degree of importance it has in their Umwelt.

Lastly, moving on to the subject of complex content; it is measured from the standpoint of the degree of cognitive elaboration required to exist in the individual's mental world, and even more so to be expressed. As a direct link to my own research, emotions find themselves at the forefront, as the simple fact that they exist entails extremely strong social ties, or a high-performance long-term memory, or a theory of the mind – and many other things. But they are far from alone in his category and I will do a more focused study of this point in the following chapters.

1.1.2 About the Corpus

What I mean by corpus involves two categories: the works in which we will find methodological tools to offer a model, and those in which we will find examples and case studies. In the first category, the works of theorists obviously take centre stage, and Peirce is most likely the one that led the way for me to build the foundations of my methodology.

The fundamental pillar of Peirce's theories is that "we have no power of thinking without signs" (Peirce 1987, p. 69) and that any thought, no matter its form, always ends up producing signs that can be detected and interpreted. This theory is explained more precisely as such:

> Les seuls cas de pensée que nous pouvons trouver sont des cas de pensée par signes. Il est clair qu'aucune autre pensée ne peut être prouvée par des faits externes. Mais nous avons vu que c'est seulement grâce aux faits externes que l'on peut connaître la pensée. La seule pensée donc qui soit connaissable est la pensée par signes. Mais la pensée inconnaissable n'existe pas. Toute pensée doit donc nécessairement exister dans des signes. (Peirce 1971 p. 66)
>
> Translation:
>
> If we seek the light of external facts, the only cases of thought which we can find are of thought in signs. Plainly, no other thought can be evidenced by external facts. But we have seen that only by external facts can thought be known at all. The only thought, then, which can possibly be cognized is thought in signs. But thought which cannot be cognized does not exist. All thought, therefore, must necessarily be in signs. (Peirce 1868a, pp. 103–104)

With this in mind, Peirce's theories are primordial for my study because of what they entail: if any thought at all (an idea, a reasoning, or the one that interests us most, an emotion) appears in an individual, there must be an observable sign that can be studied. This makes Peirce very close to the Greek notion of semeion, as explained by Favareau (Favareau 2006, p. 5):

> [...] a close association between something present and observable (pallor, a rash, a swelling of the ankles) with something else that is present and unobservable (an organ malfunction or a tapeworm within) or with something else that is non-present and unobservable (overindulgence of food and drink, an accident that happened previously) imbue this notion of **semeion** [...]

Peirce's theories also have the large benefit of making researchers try and think outside the box, outside of an experiment in which all parameters are perfectly known and all results are perfectly reliable:

> [La vérité] dépend essentiellement de ce que l'on ne professe pas que [la proposition énoncée] est exactement vraie. Mais l'on espère qu'avec les progrès de la science, son erreur diminuera indéfiniment comme l'erreur de 3,1416, la valeur donnée à π, diminuera indéfiniment au fur et à mesure que le calcul augmentera le nombre de ses décimales. (Peirce 1971 p. 89)
>
> Translation:
>
> [Truth] essentially depends upon that proposition's not professing to be exactly true. But we hope that in the progress of science its error will indefinitely diminish, just as the error of 3,14159, the value given for π, will indefinitely diminish as the calculation is carried to more and more places of decimals. (Peirce 1901, p. 718)

It is a necessary stance to take in a study like my own, that requires exploring the road less travelled.

Amongst the other works in the linguistic and semiotic fields, those regarding the notion of relevance are an important part of the methodology of my study. This notion is somewhat different from the day-to-day use of the term, but stems from it relatively directly. Relevance, as it is commonly used, is the quality of what is logical and perfectly appropriate. However, this notion in the linguistic sense of the word has larger implications:

> On pourrait comparer le concept de pertinence avec des concepts comme ceux de productivité ou de rendements, concepts qui s'analysent en termes de coûts et de bénéfices. [...] Les processus mentaux, comme tous les processus biologiques, demandent un certain effort, une certaine dépense d'énergie. [...] L'effort de traitement est un facteur négatif : toutes choses étant égales par ailleurs, plus l'effort de traitement est grand, plus la pertinence est faible. (Sperber et Wilson 1989 p. 189)
>
> Translation:
>
> The concept of relevance could be compared to other concepts like productivity or output; concepts that are analysed in terms of cost and benefit. [...] The mental processes, just like their biological counterparts, require a certain amount of effort and energy [...] The handling effort is a negative factor: all things being otherwise equal, the bigger the handling effort, the weaker the relevance.

The role of relevance theories is not to describe what happens when an individual learns or thinks or expresses themselves. Rather, their role is to explain how these mechanisms work and why a specific cognitive or linguistic "strategy" was chosen over another.

These theories can be particularly used to determine the potential reasons behind a failed test or unexpected results.

They will be necessary, first and foremost, to evaluate the degree of ecology from the experiment reports I will be analysing, and to detect any potential bias that could influence the conclusions drawn from these experiments.

Secondly, I will use relevance theories to better take into account the different parameters that come into play in our theoretical model. Is the studied species individualistic or social? Does it have a dominant sense, a privileged information channel? What do we know about its life expectancy, the age at which its members become adults, about their ability, or inability, to learn throughout their existence? These questions that are seemingly unrelated to our main topic are actually just what will form a major aspect of the model: its ecology. It is easy to understand why it is more relevant for a species to develop an elaborate mode of communication and tightly-knit social ties if it is a social species, rather than one with members that will spend most of their time isolated from one another. Similarly, understanding which is the dominant sense or the privileged information channel can explain the origin or preference for a specific mode of communication (for example, what would be the point for an aquatic species to develop an articulated language?). Finally, the length of a species' life expectancy can result in the creation of emotional links of different strengths.

Some of the semiotic works I will be using as tools belong more specifically to the field of biosemiotics. The founders of biosemiotics imagined it from the start as taking into account the elements of biology (from all species, humans included) in the semiosis process. Its birth laid the groundwork for the creation of, obviously, zoosemiotics. Their place and aspects are now considered by biologists as elements that can be ignored no longer (Boeckx and Grohmann 2013).

The theories of biosemiotics opened a path for zoosemiotics because they established the sign and its origins as primarily a biology issue:

> According to Nesher, semiosis is to be understood "as encompassing all mental processes in distinction from the physical processes" (Nesher 1990). Therefore, the problem of the origin of signs is purely biological, and is in fact one of the most fundamental problem in biology. (Kull 1992 p. 222)

In that way, they re-introduced Peirce's theories as the best theoretical system to allow viewing semiotics under its most complex form, and the aspect closest to its biological origin:

> In the light of categorical perception, icon and symbol (just like simple biological and complex human communication) become two ends of one and the same line. (Stjernfelt 1992 p. 445)

Some of the tools we will be using will, of course, come from the life sciences, mainly biology and ethology. It is not always easy for a researcher to get acquainted with material from an unfamiliar field. I will get back to this point in more detail later on, but it is part of the origins of my work on intertheoricity and the difficulty in dialogue and understanding between disciplines. A semiotician that wishes to apply his work to animals must first familiarise themselves with the theories, lexicon and methods of the field. In my case, my entry point into ethological research were the works of Bickle (2009), Danchin et al. (2005), Machamer (2009) and in order to learn the history of the field, Darwin (1872, ed. 1989).

The second category – the works in which I will find examples and case studies – is unique due to its multimodal aspect; it draws from experiment reports (including text document) but also from scientific recordings or what I will consider "wild" videos. I deem this aspect of the corpus worthy of its own discussion, which will happen below, in the part title *Including new corpus categories*.

1.1.3 About the Academic and Social Impact

I did not choose to begin this methodology proposition with a long discussion about the definitions and concepts we will be using in order to stand out from an academic level.

This endeavour is simply because it seemed problematic not to strongly question what had been done before due to the troubled past relationship between linguistics and animal studies. Much like Astrid Guillaume (2014) did in her works in semantics and semiotics, which she began by verifying the definitions of keywords that push aside animal behaviour, it seems essential in animal studies to define, or maybe even redefine certain words in the light of scientific progress. For example, specific reasons why language – and its study – should be "exclusive to human communications" (Neveu 2004, pp. 173–174) are nowhere to be found. Complex communication systems exist in large numbers in the animal kingdom, as does the use of sign systems. It might be argued that humans are the only species that use complex communication systems that involve vocal signs, but not only is that very difficult to be certain of in this field, it also has the side-effect of considering that non-vocal languages like sign languages (ASL, etc) are not actual languages. It remains nonetheless to this day one of the most frequent reasons put forward by academic circles to dismiss animal studies from linguistic and semiotic research.

However, some voices in language sciences are starting to be heard when asking for a rework of dictionary definitions that omit the topic of animals, or a study on what the specifics of human languages might be compared to animal languages, without denying the latter's existence (Agha 1997). It is from research resulting from this school of thought that I will rethink the definitions and concepts in the following chapters.

This work on definitions and concepts was also an opportunity during my doctorate thesis to "introduce" to each other the different disciplinary fields we would have work together. That step allowed me to question the definitions and concepts in fields I were not specialised in (particularly ethology) from a layperson's point of view. From this relatively "naïve" questioning came a problem associated with a basic concept: the stimulus-response coupling. The issue being that it favoured as an explanation to any and all behaviour over any other possible answer, which left no room for semiotic research. A biology school of thought, behaviourism, followed that reasoning to consider that every animal action was just instinctive and that it was therefore pointless to search for any cognitive process, mental world or motivation.

This opinion, however, is far from universally agreed upon in the scientific community, and has started to make itself scarce. For example, we can quote Peter Machamer (2009) on the matter:

> There is nothing wrong per se with behaviourist [...] experiments and models. Indeed, it might be plausibly argued that is in some cases, especially neurobiological cases, this is all we are able to do. [...] What is assumed is much more tendentious, that such experiments and models are all we need.

While it is completely understandable that those methods remain the only ones we have to study certain aspects of the animal kingdom, the amount of progress we have made in primatology, neuroscience and other fields makes it strange that we do not ask ourselves more questions about the lives of species we regularly learn have more in common with us than previously thought. More and more researchers in fields associated with those questions lament the lack of "something else" – other tools, other models, other methods, other approaches, other points of view on the subject – that could help answering them.

Defenders of behaviourism continue to believe it is one of the few schools of thought that prevent researchers from falling into the trap of anthropomorphism. However, I deem that this assertion is becoming an ideological position more than a scientific truth – and later on in this book, I will come back to the special case that is anthropomorphism in semiotics applied to animals. To me, as a result, it constitutes a large experimental bias from the researcher themselves put in place before any testing protocol or scientific model, rather than the proof of objectivity that the behaviourist methodology is supposed to be.

The idea of a biology approach in semiotics, or a semiotics approach in biology, will slowly grow into a scientific discipline that will be applied to all species (ours included) (Sebeok and Umiker-Sebeok, 1992).

The point of it – both in general and in my study itself – is its ability to reconcile both fields into a shared task through its object of study – a semiotic manifestation in a biology framework:

> Zoosemiotics intended as a mediating concept for reconciling these two seemingly antithetical spheres of discourse, ethology and semiotics: the former, anchored in the realm of Nature, embracing the totality of the multifarious phenomena of animal behavior on the one hand, the second, rooted in the matrix of Culture, traditionally held by many to comprise exclusively man's signifying competence on the other. (Sebeok 1976 p. 86)

This ambition to have fields cooperate is stated early on (we will return to this later) and explains why zoosemiotics had a better research affinity with certain schools of thought than others. It is still, without a doubt, possible to find strong detractors of the discipline to this day.

However, it is now impossible to deny that it has a place in scientific research, especially when it comes to particularly intelligent and complex species. It would be hard for us to know if Premack, the Gardners, or Fouts had any knowledge of the works of Sebeok and other founders of zoosemiotics when they began working. If they did, there is no doubt that those zoosemiotics works contained concepts that echoed their own. If they did not, it is impossible not to note the convergence between study reports and a terminology that, until then, only belonged to linguistics, such as the use of terms like "interlocutors", "signs" or "conversational partners" (Bodamer and Gardner, 2002):

> The Cross-fosterlings developed into conversational partners because familiar human interlocutors had always treated them as conversational partners. Interactive sign language had always been an integral part of their lives, beginning at an infantile level and rising to gradually more sophisticated levels as they matured.

This questioning, which started with a methodological goal, has helped me comprehend the impact and reception that a work of animal semiotics could have in academic circles. Therefore, it was logical to wonder about a potential societal impact, especially since the subject of animals has been met with a popular passionate response in recent years. However, it can sometimes work against the scientific community's image. Its credibility on the subject has been challenged, with some good reasons, due to being accused of putting the subject aside for so long. The public's distrust tells us two things.

First, it tells is that we should be wary of any and all ideological positions within science. On this note, all disciplines would benefit from the removal of barriers between academic fields: pairing language sciences with life sciences, linguistics with philosophy or semiotics with neurosciences could all prove to be extremely fruitful, and the use of different approaches of a particular subject, and science in general, help avoiding a field's withdrawal into one specific ideology, which, in the long term, can stop it from evolving and improving.

The second thing that I think is important to note here is that we have probably far too much neglected the popularisation of science under the pretence that it was not as "noble" a pursuit as scientific publication. But we do not produce knowledge, discoveries and skills solely for the scientific community, and the public also thirsts for knowledge. We only need to look at the large and sudden growth in popularity of channels that deal in popularisation of biology, physics, history or mathematics on online social platforms. Scientific jargon, while often necessary, not only slows interdisciplinary communication, it also has a similar effect between the scientific community and the rest of the population. The sudden interest in the topic of animals is the perfect opportunity to create a link that is quite marginal, for now, but ever more necessary. When civil society seizes a question that it deems important (here we find ecological, judicial, spiritual and cultural implications – we will discuss it in the subchapter about ethics), we must give it the best possible elements to try and answer it. If we do not, others might do it for us – with all the twists it can entail. This line of thinking has been the source of another decision regarding our methodology: using a writing style that is as accessible as possible, using little jargon and clarifying it when necessary.

1.2 Why Use Semiotics in Animal Studies

Semiotics as seen by Sebeok and his questioning regarding the animal kingdom led to a need to call into question a number of theories, assumptions and definitions. The question of knowing whether a language absolutely needed to be vocal to be seen as a language at all, of whether it was a faculty that comes from nature or nurture, of whether the arbitrariness of the sign could be tested in a species other than us raised many more questions than it solved and led the way to a new field of linguistics.

However, one of the distinctive characteristics of this semiotics (that we have embraced, as mentioned earlier) is that it straddles two disciplinary fields. Kull (2014, p.47) defines it as such:

> Zoosemiotics is semiotic zoology – a semiotic approach to the study of animal life. More precisely, it is appropriate to define zoosemiotics as the study of animal semiosis.

Therefore, it seemed wise to also present the discipline's subjects from the point of view of life sciences in general, and ethology in particular.

I will start by a brief summary of its history amongst that scientific field, of the various schools of thought that came from it – giving me the opportunity to point out the ones I agree with and why – as well as the area of study that the discipline currently occupies.

It is not the point of my study to retrace the entirety of the discipline's history and its twists and turns, and my presentation will confine itself to the schools of thought I will be drawing from, the unanswered questions I will be approaching, how I stand on those questions and why, as well as the impact of all of this on my study.

I will finish with a more detailed explanation on the necessity to link scientific disciplines in a work of semiotics applied to life sciences despite all the challenges it entails. While it may seem superfluous at first, it actually is an absolutely fundamental point of my study.

1.2.1 History of Semiotics

Semiotics comes from the Greek semeion, a branch of medical science that uses exterior, visible signs to see the hidden face of disease. These signs can either reveal something that is present but invisible because it is inside one's body – like someone's extremely pale skin can give anaemia, or a yellowish complexion can show hepatitis – or they can reveal something that is no longer there but has left some traces – such as an old accident that still causes one to limp, or a person suffering from lung cancer coughing black phlegm, a sign of long-time smoking habits.

With this in mind, it makes perfect sense that this discipline that now falls under the umbrella of human sciences can easily adapt itself to life sciences, as it is where it truly came from. It also makes it all the more surprising that most semioticians have shown such unwillingness for so long to get themselves acquainted with these fields. Many of them; but not all. If the French semioticians like Greimas and De Saussure would heavily focus on text and would mostly ignore the other modes of expressions (there are however some works from Greimas on the semiotics of passion) (Greimas 1981, 1988), this was not the case of the semiotician I will mostly use, Peirce.

Peirce's works are not precisely recent or unknown, even if they are not all translated in French (Deledalle 1971) and as a result are often ignored by Gremaissian and Saussurian semioticians. But the foundations they laid are a precious help when it comes to understanding how to approach our subject.

First off, Peircean theories explain how signs can be a bridge that helps our mind comprehend another species': because signs are the primordial, most fundamental element of thought. Peirce himself asked the question in *À la recherche d'une méthode*: pouvons-nous penser sans signe? (Peirce 1993). He then answers that question in the second chapter, as a consequence of the four inabilities of humans that he had listed earlier: "We have no power of thinking without signs." (Peirce 1868b). As a result, it appears obvious that signs should be our primary tool.

Thankfully, more recent semioticians showed an interest in Peirce's works and made them many of his concepts more accessible. Eco, who often studied his works, goes further and explains that an understanding of Peircean semiotics is necessary for works of zoosemiotics:

> *Naturellement, une fois cette prémisse [l'icônisme primaire] acceptée, ces situations à mi-chemin entre l'icônisme naturel primaire et les systèmes cognitifs non-humains, tels les cas de reconnaissance et de mimétisme chez les animaux (cheval de bataille […] des chercheurs en zoosémiotique) peuvent alors être abordés.* (Eco 1999 p. 29)
>
> Translation:
>
> Naturally, once this premise [primary iconism] is accepted, these situations that are halfway between natural primary iconism and non-human cognitive systems, such as cases of recognition and camouflage amongst animals (a hobby horse […] for zoosemioticians) can be tackled.

Another necessary aspect of my work that also stems from Peircean semiotics that is also discussed by Eco in the same book is the idea of thinking about the process at the origin of sign production (which takes us back to my arguments about the notion of saying):

> *[La sémiotique] doit non seulement se demander « à quoi nous référons-nous lorsque nous parlons et quel crédit lui accordons-nous ? » […], mais encore, « Qu'est-ce qui nous fait parler ? »* (Eco 1999 p. 19)
>
> Translation:
>
> [Semiotics] must not only ask "what are we referring to when we speak, and what credit do we give it?" but also "what makes us speak?"

While the two previous aspects were theoretical and methodological, the third one that will be particularly useful to us can almost be described as ideological. Peircean semiotics has always been able to adapt to the different academic fields it was introduced into. Here, I use it in the context of animal studies, but I am not the first to introduce its terminology, theoretical framework and methodological pragmatism into disciplines that are more or less distant from language sciences (such as cryptology). Therefore, Peircean semiotics seems very well adapted to researchers that worry about restricting themselves to a narrow frame that harms their study subject, and also ensures progressing within a methodological frame that stays coherent and rigorous.

All three of these aspects are necessary for semiotics as I intend to use it in this type of work. A sign is a primary and fundamental element of any form of language, thought or complex communication (the presence of a sign will not automatically entail the presence of a language, other elements will, of course, have be taken into account), and as a result the sign is the element that we have to focus on to study these manifestations as best we can.

It is also the end product of a thought, a cognitive process, and as such has a reason for being, a justification in the subject's mental world (which can be completely involuntary or unconscious, most likely in particular when it comes to emotional justifications). If Eco did not want to enter what he called the "black box"[1] in his works, it is an attempt we have to make to comprehend emotions and their semiotic manifestations.

In later years, Eco worked with someone that may be seen as the founder of biosemiotics, Thomas Sebeok. With him, a new era of semiotics begins, one focused on having the discipline interact with other sciences, particularly ethology for the zoosemiotic component. It is difficult to glean from the works Sebeok has left us what he thought of applying semiotics to animal studies. A glance at the table of contents of his 1968 book (Sebeok 1968) tells us that out of 24 chapters, only nine dealt with animal communication (some of which had very vast themes such as "land mammals") when it is clearly the given theme of the book, when six are fully dedicated to comparisons with humans.

It is possible that a large part of the difficulties animal semiotics studies comes from the fact that they focus mainly on comparisons with humans in order to find what may be "missing" from other species or to what extent they are "like us",

---

[1] *"Dire que le cognitivisme s'interroge sur les chats et les chaises ne veut cependant pas dire que les conclusions auxquelles il soit arrivé (qui sont nombreuses et discordantes) soient encore satisfaisantes. Le fantôme du schématisme hante de nombreuses recherches contemporaines, mais le mystère de cet art secret n'a pas encore été dévoilé. Je ne prétends pas le dévoiler dans ces pages, et, comme nous le verrons, je voudrais même éviter de mettre le nez dans la boîte noire de nos processus mentaux ou cérébraux."* (Eco 1999 p. 126)

Translation: "To say that cognitivism asks questions about cats and chairs does not mean that the conclusions it comes to (as numerous and discordant as they are) are still satisfactory. The ghost of schematism haunts much of contemporary research, but the mystery of this secret art has yet to be discovered. I do not have the pretension to reveal it in these pages, and as we will see, I would even like to avoid digging into the black box of our mental or cerebral processes" instead of trying to find, for example, a descriptive semiotics of the various forms of expressions found and what could be concluded from those observations.

There is no doubt, however, that Sebeok very much opened a path for works such as my own, especially by showing that semiotics was particularly adapted to this type of studies if they ever were to be carried out. Amongst the many reasons that can justify the use of semiotics in animal studies, there is the substantial amount of analytic tools at its disposal and the various degrees of mental, emotional, language complexity and sophistication they can report. We can add, as does Sebeok (1973, p. 40), the use that semioticians have of cryptanalysis:

> It has been pointed out that a linguist approaching a totally unknown language acts as a cryptanalyst: he receives messages not destined for him and not knowing their codes. The student of animal communication also resembles a cryptanalyst.

1.2.2 History of Animal Studies

The very word "animal" comes from Latin, where the word "anima" was the word used to talk about the life, the vital principle, but soon was used to talk about the soul (in French, "anima" became "âme", the word for soul). Then, there was a separation between the animal "the life being" and the human and its "âme", "the soul being".

The idea was widely used by philosopher René Descartes and his follower Malebranche. To talk about animals, he created the concept of "animal machine". The meaning was that animals were like clocks: moving, making noises, with delicate pieces and even a beat, but without soul, ideas or emotions.

This idea was very encouraged by the global occidental culture he (and we) lived in: a monotheist culture, with Man, made in the image of God, above every other creature. This lead to a conflict between reason and emotions, in which the animal seemed to always be losing. One time, the noble emotions, like altruism, were the proof of our superiority, and animals were presented without emotions, another time emotions were the animal part of us, like anger, and we were superior because of our reason, which animals could not have. Today, we still have some researchers trying to find out what is the "*propre de l'homme*": once, it was the empathy, then the ability to anticipate, then tool building etc. It changed every time another scientist discovered that some animal was doing it too.

Any work that would tend to erase the frontiers between man and animal would be widely regarded as anthropomorphism, which has sometimes led to grave methodological mistakes, as Boesch demonstrated in his meta-analysis of comparative studies between humans and monkeys:

> *Lorsqu'une barrière physique séparait l'expérimentateur du singe (mais pas de l'humain testé de façon comparée), dans treize cas sur seize, il fut noté une supériorité de notre espèce, tandis que dans les sept autres (sans barrière physique) une seule aboutit à ce résultat. La présence d'un parent se traduit par le même déséquilibre, et si l'on prend en considération la somme de ces deux biais (barrière physique et absence parentale), on voit*

que les dix études dans lesquelles ils sont présents concluent à la supériorité humaine, contrairement aux cinq autres dans lesquelles ils sont absents. (Boesch 2007)

Translation:

Significantly more studies that used physical barriers for chimpanzees and not for humans concluded that humans were superior to chimpanzees than those studies avoiding the use of physical barriers (barrier: 13 of 16 studies; no barrier: 1 of 7 studies; Fisher exact test: p=.0049). Similarly, significantly more studies that allowed the presence of one parent only for human children concluded that humans were superior to chimpanzees than studies that avoided the presence of parents during the testing (parent present: 11 of 12 studies; no parent: 3 of 11 studies; Fisher exact test: p=.0028). Finally, all comparative studies that used physical barriers and allowed the presence of one parent for children concluded that humans were superior (n10), whereas none of the studies avoiding both did so (n5; Fisher exact test: p=.0003).

To Boesch, these errors were mostly due to a major difference in treatment between the human subject and the simian subject, caused by a fear of being labelled as anthropomorphism if the experimenter were to take into account the factors of anxiety and discomfort in the monkeys. According to him, many experiments tried so hard to avoid an accusation of anthropomorphism that they started being guilty of anthropocentrism.

Since I discussed it as a conduct to avoid earlier, it is pertinent to ask what exactly anthropocentrism is and how it can be problematic in works such as ours.

Anthropocentrism is a philosophical and methodological operation that considers humans as the one reference point from which all ability, process or behaviour must be evaluated. On the opposite side, we can find anthropomorphism, which is the operation of systematically attributing human abilities or motivations to animals (anthropomorphism can also be extended to plants and inanimate objects).

Both are insidiously present in every work on the subject of animals. In the case of the former, it is because it is easier for us to judge by using ourselves as a comparison, and for the latter, it is because it always seems obvious for us to put ourselves "in someone else's place" to explain an operation or a motivation. Both stem from the same issue: how do we study and understand what is fundamentally foreign to us?

Both are necessary. Anthropocentrism allows us to think in terms of specificity of this or that human ability and can lead towards scientific thought processes: we have this, or we can do that, where does it come from, is it also in other species? Anthropomorphism lets us find answers for phenomena that are difficult to observe like thought or motivation: we think or feel or reason that way, that ability has to come from somewhere, it must have a source in the evolutionary tree, so other species can also have at least some of those abilities.

Both are obstacles. Anthropocentrism has gravely harmed animal language studies, as mentioned earlier, and still cripples the discussion on the topic to this day. The never-ending need to find a "*propre de l'Homme*" ("proper to Man") has often led to the idea that language fit the bill (Lévi-Strauss, 1958) and that nothing that did not correspond perfectly to the criteria used to describe human languages could be called a "language". Anthropocentrism was also the origin of the doctrine that says "at no point should an animal activity be interpreted in terms of complex psychological process if it can be interpreted as a less elaborate process in the scale of evolution and psychological development" (Morgan, 1903). This principle, later known as "Morgan's canon" was meant to avoid the trap of anthropomorphism, which is to say seeing an animal behaviour as much more complex than it actually is.

However, as Frans de Waal (2013, p.50) explains, the main trap of anthropocentrism has been that researchers have taken so many precautions not to be accused of anthropomorphism by their peers that they went far too excessively in the opposite direction. Behaviourism explained away everything non-human as "reflex" or "instinct", eventually explaining that humans were not exempt from this and were incapable of altruism, making the difference between good and bad, and that they were "selfish and cruel" (*Ibid.* p. 60). That is what we called the veneer theory. Ghiselin, a fervent defender of it, once told de Waal "scratch an altruist and watch a hypocrite bleed" (*Ibid.* p. 49). These were dominant theories until, ironically, animal research in the 2000s showed that altruism and moral sense in humans are instinctive processes (Haidt 2001).

If I made the choice in this work to be particularly wary of anthropocentrism, it is because I believe it has done a great deal of harm to animal research, no matter the field. Linguists considered it a non-question, primatologists rejected Goodall's works (despite the fact that she is now held in the highest regard by her peers) because she gave names to apes she studied instead of numbers (Goodall 1998), Fouts was long silenced, and the public is only now learning about works that are sometimes over 40 years old. Dario Martinelli most likely feels similarly when he writes:

> Anthropomorphism is considered, in all respects and without exception, a scientific mistake, which, historically, seems to have damaged or delayed fair interpretation of animal behavior. […] It seems rather clear that the attitude emerging from this definition is of a binary and qualitatively anthropocentric type. (Martinelli 2006 p. 487)

The linguistic, cognitive and emotional abilities of our species are inordinately complex and it is impossible to pretend nowadays that they spontaneously appeared in the infinitesimal amount of time that followed the separation of our branch from the *Pan*[2] branch – and I am only mentioning the most recent separation. It seemed logical to consider this work, not as having to place a barrier between our species and the others, but rather as trying to understand how they are related to us and what form this continuity can take.

At one time, this continuity was highly searched for in language, when we thought acculturated chimpanzees could develop a spoken language. After the failures of these first projects came ones that aimed to teach sign language to monkeys; that is the wave that the Gardners, Fouts, Patterson, Premack and Savage-Rumbaugh belonged to. While their works are now well-documented and recognised, there is

[2] The *Pan* genus contains the species of *Pan paniscus* (bonobos) and *Pan troglodytes* (chimpanzees, four sub-species). It is the only other member of the *Hominini* tribe with the *Homo* genus (humans).

one case that is particularly telling when it comes to the state of mind of researchers on the topic: Terrace's Nim project.

I could spend a lot of time discussing Nim (Terrace 1987), who was reported as unable to produce real sign sequences (which was indeed the case) without taking into account that he had learned American Sign Language in a completely rigid environment, with no real social ties, through methods that were closer to computer programming than educational methods (Forsberg et al. 2012). This case alone continues to maintain a certain amount of doubt regarding the language capabilities of great apes, while in the meantime, the subjects of Fouts, Patterson, Savage-Rumbaugh or Premack have had remarkable performances over the years (Morin 1974). Incorporating this is avoiding the trap of anthropocentrism. Just because these results are troubling to us does not mean we should not admit they are the closest thing we have to reality today. However, the debate had the merit of introducing the language sciences again into animal studies, as the success or failure of this or that project came under the responsibility of linguistics and semiotics.

Nowadays, animal studies face different problems, caused on one hand by the new sensitivity expressed by civil society about the subject of animals, and on the other, that many current topics had barely been discussed in the previous century, with animal well-being as the most striking example. Now more than ever, animal studies must evolve towards pluridisciplinary practices, capable of taking into account every challenge and force. As a result, I believe that a work of semiotics methodology in animal studies such as this one has a perfect timing.

1.3 Questions About Methodology

The challenge of research such as this one is that it requires the collaboration of various scientific fields that each have their own theoretical frameworks, methods, definitions and notions. As a result, I will fragment my presentation of this methodological framework in different parts by theme so as to make it as clear and accurate as possible.

The various debates and controversies that the study of animals and their potential semiosis have been at the centre of will be discussed, so as to trace the history of those issues and avoid any future controversy. Those debates strongly focused on the definitions on animals and their study that have been either kept or rejected, and as a result we will dedicate a chapter to the lexicology of the field. I will present the semiotic tools that were retained and how to arrange them in a larger model that will be able to coexist with life sciences, particularly ethology. I will also be dealing with the problems that animal studies are facing nowadays, especially ethics problems, and how the building of a proper zoosemiotics may be the solution.

However, before we get to those chapters, a methodological approach such as this immediately raises two issues that I have to address: the issue of the multi-field approach, and the issue of the multimodal approach.

It is indeed complex to propose a research framework that relies on different fields when the very way university trains people nowadays is to make them specialists of one specific field. This is a primary issue in my research and my process, and I chose to use the concepts, studies and works that do not belong in our field with the required utmost carefulness. This means a need to systematically justify any choice being made in regards to any study or methodology used instead of another, a reliance on what meets the general scientific consensus or has been confirmed by peers, and only proposing another use or interpretation of data if done with a sound concept that can cross the bridge between disciplines. These cases will be given a point-by-point explanation in following chapters whenever necessary. I have put this study together in a way that allows non-specialists of linguistics to use it with as little difficulty as possible, hence the great precision on our part about concepts, schools of thought and methods. I realise that this way of working may feel slightly repetitive for specialists, but I believe making research accessible is now a bigger priority than conciseness.

Different research fields mean different work methods and different types of corpus. Where linguistics has a large percentage of its corpus being texts, ethology has a large part of multimedia content, be it experiment recordings, subject photographs or field videos. Therefore, part of my work will be to make objects of widely different types coexist inside a mixed corpus, to give each of those types some selection and rejection criteria to preserve the scientific integrity of my corpus, and to properly set into place each object where it belongs in my work (texts in the theoretical framework or the setting up of the model, experiment recordings will be used for the ecological validation of a concept or the weighing of part of the model, etc). Each individual case will have a detailed explanation if deemed necessary.

I will also take a look at the very specific way interdisciplinary methodology functions and the challenges it deals with in terms of premises, academic traditions, and also the specific advantages it contains and why it seems like a promising research lead. This point will also allow me to justify part of my methodological working and my stances.

1.3.1 Studying a Subject from Different Academic Fields

One of the first challenges stems from the fact that, to work in this discipline, one must have a certain understanding of both the "zoo" part and the "semio" part of the field. However, a full mastery of both those aspects – that is to say, biology and linguistics – seems largely impossible. It must always be kept in mind that research on animal semiotics always has the particularity of being the study of an object from one academic field with tools from another.

I have personally circumvented part of this challenge by using as research material studies that were done and reported by specialists of their own field, rather than my own observations. By collecting the results of researchers that were very often reviewed by their peers, I minimised the error margin created by my lack of specialisation in biology, and particularly ethology.

Nonetheless, a certain amount of theoretical and methodological knowledge of both fields is still required to use and understand their major components, which calls for dedication towards interdisciplinarity and a substantial amount of plasticity in definition organisation and the elaboration of methods.

While that is a serious challenge that demands a large amount of time, energy and multidisciplinary work, it also is a major asset of the method. First off, because it shows that the researcher put in a large amount of work, as it entails a will to learn about other fields and an initiative to study anything within them that could prove useful to the research subject.

Through a knock-on effect – and this is a point I have already partially brought up – tackling works from other disciplines leads to thinking differently about the way to write and organise your own works. While it is essential that specialists from our field understand our works, the issue of accessibility to non-specialists – including non-specialists from scientific circles – is rarely discussed. The difficulties a researcher meets in zoosemiotics while working with other fields are an important push towards being aware of their own work and how to make it more accessible to both the general audience and researchers from other fields.

Similarly, being confronted by other disciplines with other concepts, other theories, other ideologies and another history leads a researcher to "think outside the box" of his own discipline, to look at certain phenomena from a different perspective. This idea is made very clear by Philip Lieberman (2013, p.93) in his book *The Unpredictable Species*:

> It is clear that we can think outside the box of language. [...] Words are only one of the forms by which we express our thoughts. Linguists often seem to think that language is the only mode by which we think.

Lieberman criticises a certain rigidity that a number of linguistic theories that seem to make language the indispensable condition for thought to exist – which, in my case, is the reason why I preferred Peircean works to more recent linguistics; there is no idea without a sign, but there can be ideas with language – but many biologists or neuroscientists have similar criticism about some of their schools of thoughts that are deemed too stiff (Machamer, 2009 p.175).

It is because of the various points I have brought up here that there is a very real necessity of a good cooperation between fields to try and find solutions to challenges facing zoosemiotics and preserve its very specific assets.

This interdisciplinary cooperation not only allows, but also sometimes requires as functioning conditions some notable improvements in various aspects that are difficult to acquire in a single discipline.

I have discussed the writing aspect and the research accessibility, as zoosemiotics requires that a highly precise and clear communication be possible amongst the disciplines called in for a specific research.

I have quoted Lieberman for the methodological aspect, and the obligation to think outside the box of one's original field to get to a system that takes into account all fields involved while also being viable and functional for each of them.

But those are points that are relatively obvious once they are discussed. I would like to get back to the cooperation of knowledge and theories of each field which necessarily leads to intertheoricity. I believe it is the one aspect of interdisciplinarity that is the most likely to be disputed. I am obviously not trying to say that the involved fields are lacking if left on their own. Biology has made great discoveries without the input of human sciences, and linguistics has revealed many concepts without the help of life sciences.

However – and this is a problem that is far from specific to our subject – when a study subject is astride two disciplines, is it necessary to work with both aspects together to avoid the risk of committing grave mistakes. In other fields, that is exactly what oncology did by studying the impact of the patient's mental state on their survival chances, and what physics did by adding models of phenomena that can only be explained through quantum physics.

In the case of the fields we are dealing with, and the question we are dealing with, I can tell that many interpretation mistakes or skewed results could have been avoided by following that principle. I quoted the famous example of Benveniste and his study of the dancing of the bees, which illustrates well the biases that linguists wishing to study animal language can fall into by not looking at the works of biologists on the topic. Benveniste certainly has the excuse of working in his time and not having access to the research we have now, but even back then, a simple look at the works of primatologists would have called into question many of his conclusions, and looking into entomology would have shown that collective intelligence is a case that is very specific to some species and not representative enough to draw conclusions on a whole kingdom.

Some biologists have fallen into the same trap. Many primatologists (mainly Roger Fouts, but also Frans de Waal more recently) put Terrace's project Nim's failure down to a lack of knowledge on the nature or language, on how it functions in an individual and its building process. Forsberg and his colleagues (Forsberg et al. 2012 p. 21) go as far as claiming this failure could have been avoided, had Terrace and his team read the works of Wittgenstein on the topic:

> Observe that Project Nim applied the same metaphysical thinking against which Nietzsche and Wittgenstein were in rebellion. An idealization of human language as vocabulary and grammar was placed at the beginning of the speech, as its foundation. The assumption was that in order to speak with an ape, you must first leave life behind as a distraction and move into the more abstract domain of first mental causes (a domain which, I repeat, was the nonhuman for Nietzsche and Wittgenstein).

As a counterpoint, I call back here to what we said on the Gardners' great comprehension of linguistics mechanisms.

Therefore, the interdisciplinary aspect of a work such as mine is neither counterproductive, nor superfluous; it is more than necessary in order to correctly deal with a topic which contains elements from various scientific fields. And it invites one to think about disciplines cooperation, not as an ancillary methodology point, but rather as a fundamental and vital element of a study that aims to be as thorough as possible.

1.3.2 Including New Corpus Categories

As explained earlier, the choice of a multimodal corpus is justified by the coexistence of various disciplines amongst our theoretical framework. These disciplines all have their own theoretical corpus, work documents and reference tools.

The lack of research in animal semiotics is logically coupled with a lack of associated field studies able to provide necessary date. The semiotics corpus must therefore cast a wide net so as to gather varied documents and regroup its information.

While the exploitation of scientific literature is not a major challenge, the thorough, scientific use of multimedia documents is a lot more complex, and the odds of making errors is much greater. This is the main difficulty but also one of the main assets of a multimodal corpus.

It is mostly about exploiting documents of ethological significance, some leaning very much towards zoosemiotics, and others that have yet to be exploited by scientists regardless of their field.

We must first think about scientific recordings, that is to say images that come directly from research teams and presented to the rest of the academic world as teaching or demonstration documents. During my thesis work, images recorded by the teams of Pepperberg or Savage-Rumbaugh were part of this category.

The fact that these documents come from research teams may give the impression that they are perfectly objective and should be exploited as is without any interpretation from our part. This is not exactly correct, and while watching these images, we need to ask ourselves a number of questions before drawing any generalizable conclusions.

First, it is possible that there is an ideological orientation that has caused the publication of these images – in this case, scientific recordings are no different than other types of documentary footage. It may be interesting to try and find other audio-visual material the researchers have worked on (such as documentaries), where their ideological stance may be more obvious, to have an idea of what that stance might be or how it might have influenced the choice of images shown – or not shown.

We must also pay attention to elements allowing us to put the images back in their original context and identify them clearly. Apart from the spatiotemporal aspect (were there any ellipses or cuts?), it is important in the case of scientific recordings to, on one hand, use these images to identify possible biases directly induced by the researchers (like a bias in the protocol of the tests being filmed, for example), and on the other hand, to have the required knowledge to make sure we understand what we are watching – for example, in the case of monkeys communicating through American Sign Language, it is necessary to know that language or check with an interpreter who does to check the proposed translations.

Similarly, it is important to make sure to understand the terminology of the scientific specialty being used, as it can be a source of misinterpretation if not mastered or if used in a foreign language – it is well-known that the same term used in scientific communities from different countries can include different acceptations, nuances – even have a completely different meaning).

Despite these challenges, the addition of scientific recording is far from negligible for our study. They often are the most revealing demonstrations of complex experiences available, as well as the clearest, and grant us access to new and innovating methods or discoveries. They also allow us to record behaviours that were misunderstood or unexpected in a more precise way than we would have been able to with just a description, and reduce the error margin that would come from misunderstanding a little known phenomenon.

A semiotics corpus for this work can also contain what I call "wild" recordings. These are audio-visual recordings made by people that are not professionals and/or specialists. These recordings were not prepared or simply even planned and show an interesting event "on the spot", as it is happening.

This type of recording is very new. Until a few years ago, it was not possible for a layperson that witnessed an event to record it, sometimes for several minutes, and publish it *en masse* by themselves. The emergence of this category of document is the product of technological progress that gave access to the public to both phones capable of high-quality, instant recordings and an ever-more available internet with video-sharing platforms that they could contribute to.

As a result, the exploitation of this type of document in scientific work next to more academic recordings is relatively new, and must be approached with caution as it is difficult to have the necessary methodological distance.

However, I insist upon including them in this study for the informative value they represent. The gigantic amount of little-to-not exploited documents they currently form makes it so that it is not really scientifically acceptable to ignore them under the pretence that we lack the methods to analyse them from the right perspective.

It is, however, possible to use such documents if we apply a strict analytic method. First, we must identify the context of the video, starting with its time and place. This entails some research to find potential previous versions, internal elements allowing us to date or locate it, or a reuse of the recording from a reputable source (a scientific one, a journalistic one, etc.) which gives it contextual information.

Second, since this video has been filmed and published by unidentified people, it is possible that it is incomplete and that it was edited. We are not talking here about falsification (but we will right after this) but rather of modification, most likely involuntarily. While reviewing a recording that is several minutes long, the person recording it might have removed parts they deemed uninteresting, or during which nothing happens, simply because they saw no reason to keep them in the video. Therefore, we need to be careful and look for any sign of editing (jump cuts, objects that have moved, etc.) to re-establish the exact length and chronology of the video. Yet here we are only discussing "innocent" modifications, and those are usually not too hard to detect, as they are simply not hidden.

On the other hand, it is also possible for us to be faced with a voluntary modification, a doctored video. If some image manipulations can be visible, others are well hidden. This is why caution will always be required when being presented with a recording that contains absolutely extraordinary content. If no obvious manipulation is visible, and that it is not possible to precisely establish the context of the recording, it will be simply dismissed from our corpus as a cautionary measure. This is all the more necessary if the author, or authors seem to use the video to defend a specific ideological stance through the video description, the location chosen for the recording or the things they say in it or any other similar recording in which the authors are clearly identified. At this point, the suspicion of a voluntary manipulation for "propaganda" must be very high.

Finally, it is possible that some parts of the recorded event are missing from the video. This is intrinsically linked to the very type of document we are discussing. The event was not planned, the person filming might have missed the beginning or have been surprised or shocked, and the first few seconds or minutes are missing. And since the recording was not planned, the author usually is still where they were when the event happened, which is to say, they may be far away, or at a difficult angle, or an obstacle might be in their way.

1.3.3 Hypothesis, Biases and Ideologies

If, before we truly began this book it seemed important to me to discuss potential biases and ideological aspects while presenting the initial premises and work hypotheses, it is because I see those two as linked. Any scientific work, regardless of how rigorous it can be, is the result of the work of one or several researchers, and consequently, can be subjective, influenced and fallible. It seems to me that recognising this fact and taking it into account is the best way to defend ourselves against it.

It has been long established that even a simple opinion of an experimenter during a test can influence its results. To be convinced, one only needs to read the works on the placebo effect (Chamayou 1999) and the difference in its effectiveness in a patient depending on whether the experimenter knew if they were giving them real medication or a placebo – hence the existence of "double-blind" experiments. While it is not always possible to experiment in these conditions, it is always possible to study the major subjective stances of the experimenter to identify them and make sure they remain as objective as possible.

The primary source of subjectivity in an experimenter is the scientific movement they are a part of, the ideology or ideologies they believe in and sometimes defend. Everyone has an ideology, and anyone can be part of a movement (even the movement of people that do not want to be part of a movement). That is not the issue; the issue is in trying to recognise this scientific bias.

In ethology, Yves Christen explains that this bias is very obvious in the words of certain scientists, who claim to be "against anthropomorphism" without ever uttering the name of their movement: anthropocentrism. I have already discussed it earlier, it is not a problem in itself, human beings are exceptional creatures that have pushed their abilities much farther than other species, some may want to see humans as apart from them. But this ideology has to be clear and explicit, and that the reader sees the results of those researchers' work while knowing how they can be influenced. Griffin (1977) also pointed out this ideological tendency, especially the problems created for researchers that follow it:

> [...] it has seemed of fundamental importance to many scholars and philosophers to insist on a "radical difference in kind" in the terminology [...] rather than a quantitative difference in degree of elaboration. [They] have struggled to formulate objective criteria by which human language can be qualitatively distinguished from animal communication. But this task has become increasingly difficult as more and more is learned about the communication behavior of honey bees, chimpanzees, and other animals.

A second possible bias in the experimenter's position can be the initial premises of experiments and their assumptions, both of which sometimes come from the movements and ideologies that scientists belong to. Starting an experiment while having strong or narrow premises or already assuming certain results have great risks of tampering with the results. Of course, premises are required and it is normal, depending on the experimented theory, to expect certain results. This is why there is a need to call things into question.

In ethology, there is a number of researchers that refuse to accept the results of their colleagues that tend to prove the existence of animal intelligence, of an elaborate form of semiosis in animals or their ability to feel complex emotions. Those researchers insist that all of this is simply pure instinct. This is not a position that can be maintained, because it entails a complete refusal of experiment results from different researchers, different teams, in different places, at different times, on different species with different study points that still lead to results that corroborate each other. What also makes this a precarious position to hold is that the same scientists then have to give a precise definition of instinct that can explain all these factors without it applying to humans.[3]

Lastly, every researcher has had a personal, university, and/or professional journey that often is a source of inspiration, influences, ideas, and sometimes methods. Again, while this can be rewarding, our focal point is that we must remain aware of this possible influence: the ethologist that likes to hunt rabbits for Sunday lunch does not approach things the same way as an ethologist that grew up in a strictly vegetarian household.

Interestingly, another widespread preconceived notion is that only ethologists, the authorised scientist, is able to understand animals and provide proof of their behaviour, denying extremely data-rich studies from animal owners, veterinarians, or carers that are constantly in contact with animals. It was only recently that it was accepted that academic knowledge is not always above field knowledge or the addi-

---

[3] In *Abrégé d'éthologie*, J.-C. Guyomarc'h goes as far as classifying non-speaking humans with *Pongidae* to keep spoken language as a specific human characteristic, which we find doubtful. (Guyomarc'h 1980, p.168)

tions provided by other fields, but rather that they all complement each other. This complementarity is the basis upon our work rests, along with our will to offer a methodological model of semiotics to animal studies.

My hypothesis is that it is perfectly possible to work that way in order to study semiotic manifestations, that can sometimes be complex, in the animal kingdom, and that semiotic tools are sturdy and versatile enough to be used in another scientific field (in the present case, ethology), and produce convincing results.

However, the issue that my work really focuses on is the manner in which it is done, rather than whether it can be done. This point is the origin of several important sub-questions that my research work must solve by providing a positive and substantial response to this issue:

Conception sub-questions:

- What are the controversial aspects of the animal issue from a semiotic point of view that must be taken into account in this type of work? (Chap. 2: Debates and controversies)
- What definitions should we all agree upon from an interdisciplinary standpoint? (Chap. 3: Definitions: the necessary ones and the problematic ones)
- What are the semiotic tools and concepts that are interesting and desirable to work with in animal studies? (Chap. 4: Semiotic tools and concepts)

Creation sub-questions:

- How does intertheoricity help us build bigger, more efficient models in animal studies? (Chap. 5: Intertheoricity: how to build bigger models)
- What are the parts, strengths or flaws, of the pre-existing animal study methodology upon which our model must be built? (Chap. 6: Strengths and flaws of ethological and biological methodology)

Evaluation sub-questions:

- What are the problematic ethical aspects of animal studies that semiotics can answer? (Chap. 7: Animal studies, animal ethics)
- Finally, what are the stakes of a real zoosemiotics discipline, and why build it? (Chap. 8: Building zoosemiotics)

I hope this introduction has shed some light on the motivations and the goals I aim to accomplish. It highlights the genesis of my work, but also the historical difficulties that came before it and that must be dealt with before any interdisciplinary work. That is how I wanted to help the reader understand my method, and the reasons why the following chapters are in the shape and order that they are in.

References

Agha, A. (1997). "Concepts" and "communication" in evolutionary terms. *Semiotica, 116*, 189.
Baldwin, J. M. (Ed.). (1901). *Dictionary of philosophy and psychology*. London: Macmillan.

Barbieri, M. (Ed.). (2006). *Introduction to biosemiotics: The new biological synthesis*. Dordrecht: Springer.
Benveniste, E. (1966). *Problèmes de linguistique générale*. Paris: Gallimard.
Bickle, J. (Ed.). (2009). *The Oxford handbook of philosophy and neuroscience*. Oxford: Oxford University Press.
Bodamer, M., & Gardner, A. (2002). How cross-fostered chimpanzees (*Pan troglodytes*) initiate and maintain conversations. *Journal of Comparative Psychology, 116*(1), 12–26.
Boeckx, C., & Grohmann, K. K. (Eds.). (2013). *The Cambridge handbook of biolinguistics*. Cambridge: Cambridge University Press.
Boesch, C. (2007). What makes us human (*Homo sapiens*)? The challenge of cognitive cross-species comparison. *Journal of Comparative Psychology, 121*(3), 227–240.
Chamayou, G. (1999). L'essai "contre placebo" et le charlatanisme. *Pour la science, 38*, 14–17.
Danchin, E., Giraldeau, L.-A., & Cézilly, F. (Eds.). (2005). *Écologie comportementale: cours et questions de réflexion*. Paris: Dunod.
Darwin, C. (1989). *The expression of the emotions in man and animals*. London: W. Pickering.
Deledalle, G. (1971). *Le Pragmatisme*. Paris/Montréal: Bordas.
Eco, U. (1987). Notes sur la sémiotique de la réception. *Actes Sémiotiques Documents, IX*(81).
Eco, U. (1988). *Sémiotique et philosophie du langage*. Paris: Presses universitaires de France.
Eco, U. (1999). *Kant et l'ornithorynque*. Paris: Grasset.
Favareau, D. (2006). The evolutionary history of biosemiotics. In M. Barbieri (Ed.), *Introduction to biosemiotics: The new biological synthesis* (p. 5). Dordrecht: Springer.
Forsberg, N., Burley, M., & Hämäläinen, N. (Eds.). (2012). *Language, ethics and animal life*. New York: Bloomsbury.
Goodall, J. (1998). Learning from the chimpanzees: A message humans can understand. *Science, 282*, 2184–2185.
Greimas, A. (1981). De la colère : étude de sémantique lexicale. *Actes sémiotiques – Documents, 27*.
Greimas, A. (1988). De la nostalgie: étude de sémantique lexicale. In *Hommage à Bernard Pottier* (Vol. 1, pp. 343–349). Paris: Klincksieck.
Griffin, D. (1977). In T. Sebeok (Ed.), *How animals communicate* (p. 29). Bloomington: Indiana University Press.
Guillaume, A. (2014). Animal : « être sensible » unanimement désensibilisé. Sémiotique du sensible. *Revue trimestrielle de la Fondation Droit Animal, Éthique et Sciences, 81*, 35–37.
Guyomarc'h, J.-C. (1980). *Abrégé d'éthologie* (p. 1980). Paris: Masson.
Haidt, J. (2001). The emotional dog and its rational tail: A social intuitionist approach to moral judgment. *Psychological Review, 108*, 814–834.
Kull, K. (1992). Evolution and semiotics. In T. A. Sebeok, D. J. Umiker-Sebeok, & E. P. Young (Eds.), *Biosemiotics: The semiotic web 1991* (p. 222). Berlin: Mouton de Gruyter.
Kull, K. (2014). Zoosemiotics is the study of animal forms of knowing. *Semiotica, 198*, 47–60.
Lévi-Strauss, C. (1958). *Anthropologie Structurale*. Paris: Plon.
Lieberman, P. (1977). In T. Sebeok (Ed.), *How animals communicate* (p. 23). Bloomington: Indiana University Press.
Lieberman, P. (2013). *The unpredictable species*. Princeton: Princeton University Press.
Machamer, P. (2009). Learning, neuroscience, and the return of behaviorism. In J. Bickle (Ed.), *The Oxford handbook of philosophy and neuroscience* (p. 175). Oxford: Oxford University Press.
Martinelli, D. (2006). In M. Barbieri (Ed.), *Introduction to biosemiotics: The new biological synthesis* (p. 487). Dordrecht: Springer.
Morgan, C. L. (1903). *An introduction to comparative psychology* (2nd ed.). Londres: W. Scott.
Morin, E. (Ed.). (1974). *L'Unité de l'homme : invariants biologiques et universaux culturels*. Paris: Éditions du Seuil.
Neveu, F. (2004). *Dictionnaire des Sciences du Langage*. Herstal: Armand Colin.

Peirce, C. S. (1868a). Questions concerning certain faculties claimed for man. *Journal of Speculative Philosophy, 2*, 103–114.
Peirce, C. S. (1868b). Some consequences of four incapacities. *Journal of Speculative Philosophy, 2*, 140–157.
Peirce, C. S. (1901). Truth. In J. M. Baldwin (Ed.), *Dictionary of philosophy and psychology* (Vol. II, pp. 716–718). London: Macmillan.
Peirce, C. S. (1971). *Le Pragmatisme*. Paris/Montréal: Bordas.
Peirce, C. S. (1987). *Textes fondamentaux de sémiotique*. Paris: Méridiens-Klincksieck.
Peirce, C. S. (1993). *À la recherche d'une méthode*. Perpignan: Presses Universitaires de Perpignan.
Pepperberg, I. (1999). *Alex studies. Cognitive and communicative abilities of grey parrot*. Cambridge: Harvard University Press.
Sebeok, T. A. (1968). *Animal communication: Techniques of study and results of research*. Bloomington: Indiana University Press.
Sebeok, T. (1973). *Perspectives in zoosemiotics*. The Hague: Mouton.
Sebeok, T. (1976). *Studies in semiotics: Contribution to the doctrine of signs*. Bloomington: Indiana University Press.
Sebeok, T. (Ed.). (1977). *How animals communicate*. Bloomington: Indiana University Press.
Sebeok, T., & Umiker-Sebeok, D. J. (Eds.). (1992). *Biosemiotics: The semiotic web 1991*. Berlin: Mouton de Gruyter.
Smith, A. (Yankelovich, D.) (1972). *Supermoney*. New York: Popular Library.
Sperber, D., & Wilson, D. (1989). *La pertinence: communication et cognition*. Paris: les Éditions de Minuit.
Stjernfelt, F. (1992). Categorial perception as a general prerequisite to the formation of signs? On the biological range of a deep semiotic problem in Hjelmslev's as well as Peirce's semiotics. In T. A. Sebeok, D. J. Umiker-Sebeok, & E. P. Young (Eds.), *Biosemiotics: The semiotic web 1991* (p. 445). Berlin: Mouton de Gruyter.
Stoeger, A. S., Mietchen, D., Oh, S., de Silva, S., Herbst, C. T., Kwon, S., & Fitch, T. (2012). An Asian elephant imitates human speech. *Current Biology, 22*, 2144–2148.
Terrace, H. S. (1987). *Nim*. New York: Columbia University.
De Waal, F. (2013). Le Bonobo, Dieu et nous. Lonrai: Les Liens qui Libèrent.
Yin, S. (2002). A new perspective on barking in dogs (Canis familiaris). *Journal of Comparative Psychology, 116*(2), 189–193.

Chapter 2
Debates and Controversies

Abstract Conception sub-question n°1: what are the controversial aspects of the animal question from the point of view of semiotics that must be taken into account in this type of work?

This chapter will first tackle the existing controversies, by focusing on three cases that are particularly representative: controversies about language, consciousness and emotions among animal studies.

It will then present the point of view of humanities, on the question of what an animal is (and, by contrast, what a human is), but also on the question of the place and the role of language sciences. It will present the specific but largely present case of the French academic tradition in the genesis of the book. Finally, it will make explicit the position of the author concerning this book, a position characterised as "the lesser evil", and will expand more particularly on the specific treatment of the issue of definitions.

Keywords Controversies · Animal language · Animal emotions · Self-awareness · Humanities

Setting up an interdisciplinary work method involves trying to take into account all the different aspect of the studied subject (Guillaume 2014b). While they do not seem to be scientific aspects per se, the debates and controversies that a subject creates or crystallises on are an integral part of that subject and must be considered.

The controversial aspect of a subject is very telling about the way it is seen by science or the general population, and as a result, it also tells us the images and ideas it is associated with, which might, consciously or not, interfere with the research's methodology. That is why, in order to build the theoretical bases of our methodology in this chapter, I will strive to answer the following question: what are the controversial aspects of the animal question from the point of view of semiotics that must be taken into account in this type of work?

Some of these controversies are old and obvious to those knowledgeable about animal research: whether it is admitted that animals have language, if they have a form of consciousness (and how to measure it) or, more specific to our specialty, if they have emotions and how to talk about them. These are controversial aspects that

P. Delahaye, *A Semiotic Methodology for Animal Studies*, Biosemiotics 19,
https://doi.org/10.1007/978-3-030-28813-6_2

create a lot of tensions. This chapter gives us the opportunity to clarify what room I will be giving them in our methodology, and why.

If most of those controversies seem directed more towards the academic specialties dealing with animal studies, particularly ethology and biology, the field of humanities is not exempt from debates on the subject.

First, because the controversy about language obviously strongly affects linguists and questions the usefulness of language sciences: are they disciplines that focus solely on humans, or should they also study other species? Second, because different fields, from linguistics to philosophy, have tried to define what an animal is and what makes it different from a person, and some of those definitions conflicted with the results found by exact science. Lastly, in humanities, there are certain academic traditions that have more rigid positions than others. This point will be briefly address, as I have no intention to discuss the entire history of science, but as I also come from one of those problematic academic traditions in the French humanities, I thought it was an important point to explore so as to make the genesis of our endeavour easier to understand.

Once that is done, I will explain which positions I kept for this work and why. They can be summed up in two major aspects.

The first is the commitment to the “lesser evil”. Whenever a debate would occur with two opposing arguments that seemed impossible to reconcile, I would choose the one that looks the least harmful to the method I am building. The second aspect, which can appear quite surprising from a linguist, was to adopt a certain distrust of definitions: I realised that a large number of the debates I will address in this chapter were not conflicts of methods or hypotheses, but rather conflicts of definitions. In fact, this point is so important to us that it will be at the centre of the next chapter.

2.1 Existing Controversies

I will begin this chapter by focusing on the three controversial topics that I ran into the most frequently during my research, my thesis, and subsequent experiments.

The first topic is probably the most obvious for a linguistics researcher: Whether the term “language” can be applied to animals. While the answer to this has almost always been “no”, the reasons for that are not always clear, and many debates on the topic have occurred as some scientists believe this is no longer a reasonable position considering what we know on the subject.

The very manner we speak about the notions of language – both in the sense of the general function and as a tongue – shows a certain difficulty that comes with trying to move things in that direction. For example, in French linguistics dictionaries (I am using French as it is my native tongue and therefore the one I are most familiar with), we can find definitions such as these:

> Langage : Le langage est une fonction **caractéristique de l’espèce humaine** qui réside dans sa faculté de communiquer. Le langage détermine les conditions d’existence de la langue.

> Langue : Du point de vue de sa fonction, la langue est définie comme un système complexe de communication **propre aux communautés humaines**. Du point de vue de sa nature, elle est définie comme un système de signes **vocaux**. [...].(Neveu 2004 pp. 173–174)

Translation:

Language: Language is a function that is *characteristically human* which lies in their faculty to communicate. Language determines the conditions for which a tongue exists.
Tongue: From the point of view of its function, a tongue is defined as a complex system of communication *exclusive to human communities*. From the point of view of its nature, it is defined as a system of ***vocal*** signs.

As Astrid Guillaume (2014a) demonstrated in her semantics and lexicology works, the definition, and its evolution, of key words used to talk about animals is a major tension point, but also an indicator of ideological stances that can exist in concealed or unconscious ways. In the quoted example, there is no reason given to justify why language and languages are "exclusive to human communities". "Complex systems of communication" exist in species other than ours, in large numbers even, as does the use of "sign systems". It could be argued that humans we are the only species with sign systems with a vocal form, but not only is that very difficult to be certain of in this field, it also has the side-effect of considering that non-vocal languages like sign languages (ASL, etc) are not actual languages, as previously mentioned in Chap. 1. This conclusion disturbs me greatly.

Within language sciences, a certain change can be seen in the work of some researchers that aim to study, for example, the specificities of human language compared to animal language, without denying the existence of the latter (Agha 1997, p. 189). I will return to this endeavour in Chap. 3 when studying the issue created by the notion of language.

The second topic that attracts the most controversies, and one of the oldest, is without a doubt the subject of conscience. This concept is crucial to my work for two reasons.

The first reason is that I am studying semiotic manifestations, some of which are conscious, some of which are not. These two modes of emission do not carry the same meaning, so I need to be able to distinguish them, which implies a consensus on conscience in the species producing the signs.

The second reason is that conscience can be deeply involved in certain emotions – one of my main study subjects – which means we need to be able to evaluate its role and its place.

The first reason is especially important as I will be working with Peircean semiotic tools. Unlike other, more recent theories, especially in communication, Peirce's theories see – and work on proving – the sign as a proof of the presence of a conscience, specifically, an intelligent one.

> [Un signe est] tout ce qui détermine quelque chose d'autre (son interprétant) à renvoyer à un objet auquel lui-même renvoie (son objet) de la même manière, l'interprétant devenant à son tour un signe et ainsi de suite ad infinitum. Il est certain que la conscience intelligente doit entrer dans la série. (Peirce 1992 p. 126)

Translation:

[A sign is] anything which determines something else (its interpretant) to refer to an object to which itself refers (its object) in the same way, the interpretant becoming in turn a sign, and so on ad infinitum. No doubt, intelligent consciousness must enter into the series. (Peirce 1901, p. 527)

The idea that there cannot be signification without intelligent conscience, even a relatively primitive one, has enormous implications for my work. However, semioticians subsequent to Peirce have made this analysis a bit more complex by considering that it is not necessarily just the sign, but rather the act of semiosis in its entirety that must be analysed, adding a third element to the interpretant and object:

> Dans un rapport de sémiosis le stimulus est un signe qui, pour amener une réponse, doit être médiatisé par un troisième élément (nous pouvons l'appeler « interprétant », « contenu », « renvoi au code »...) qui permet au signe de représenter son objet pour le destinataire. (Eco 1984 p. 22)

Translation:

In a connection between the subject and their semiosis, the stimulus is a sign that requires a third element – we can call it "interpretant", "content", "code reference" – to receive a response. This third element represents its object to the recipient.

The second reason will be more anecdotic in this book, but it led to part of my methodology. I feel I should mention it, but that will just be done briefly in this chapter.

The last controversial topic, to which I will devote part of this chapter, is the subject we have made our specialty – emotions in non-human animals.

From a historical perspective, emotions have been viewed as a human-only concept, sometimes a gift, sometimes a plague, but always without use in nature and other animal species. Ethology inherited this ideology, and at its outset, some data or behaviours were simply not reported because they did not match this view of what emotions should be.

But quite recently, the scientific viewpoint about what an emotion is has started to change. In this part of chapter, we would like to show how this new mindset has deeply changed our perspective of emotions as well as animal life. Particularly, this change in ideology affected our understanding on some unforeseen behaviours, and we would explain how they are completely logical as soon as emotion is taken into account, but impossible to understand without this concept. We also would like to show how ethology, by starting to study emotions, is more and more forced to integrate the notion of personality, or at least, distinct individual lives when studying other species.

This part of chapter will show as well how our better understanding of the capacity of animals to experience emotion should lead us to study their emotional life in ethology the same way we already study their social or reproductive lives. We will also introduce some perspective about how this understanding can be a way into a better knowledge of evolution and phylogenetics of emotions, in our species and lineage, as well as in other animals'.

2.1.1 Language

It seemed crucial for a study focusing on signs and sign systems emitted by other members of the animal kingdom to come back to the subject of language, specifically language in the animal kingdom.

Of course, it goes without saying that semiotics does not only deal with language, but I have also noticed during my preparatory work that it was almost impossible to talk about zoosemiotics without the problem of "animal language" reappearing. This debate was particularly virulent in French speakers – I will get back to this point later on.

Every time someone puts "language" and "animal" in the same sentence, a long and sterile debate starts. Can we call this or that "language"? How to universally describe language? What is the objective definition of such a particular notion? And why is this point so important in linguistics? We tend to believe that language is the most elaborate and precise way we, human beings, have to communicate things, to understand things, to describe what happens in our lives.

The very question of whether to use the term "language" in the context of animal communication studies is an old and difficult one. It was considered several time to be "solved once and for all" by linguists of various influences in their field, Emile Benveniste (1966) being at the forefront, which in itself is fairly revealing of the real difficulty to settle the question in a definitive manner. As Favareau (2006, p. 29) suggests, maybe the question was just asked in the wrong way:

> [...] the proper question is not "Do animals have language the way that we do, and if not, why not?" but rather: "As animals ourselves, how is it that we have language?"

It seemed both prudent and necessary to begin by looking into this question before starting any scientific work. This questioning is all the more necessary because zoosemiotics is a discipline of language sciences first and foremost. This controversy is also the opportunity to clarify why our zoosemiotics work very quickly focused on the topic of emotions.

Because there is one field in which language fails us in this way. The field of emotions. If we put aside what Jakobson called the poetic function of language, which tries its best to compensate this short coming, language fails to help us express emotions. Of course, you can say "I'm sad" or "I'm glad", but these sentences don't seem enough to someone who just lost a friend or became a father. With language, any language, we can only express a fragment, a minor part, of what we feel. Emotions are the language failure. Still, emotions are a very concrete and important phenomenon, they produce signs in those who feel them, and these signs are entryways giving us a way to study them. This study is only possible with semiotic tools, because where language fails, we only have those signs to help us understand what is going on. An interesting side effect of using semiotics is the removal of the usual long and sterile debate that starts every time we ask "and, on this subject, what about other animals?" Where language seems to be a sensitive subject, the private turf of linguists, semiotics seems to be, on the contrary, a more peaceful field.

Rather than talking about language, semiotics in general and biosemiotics particularly prefer using the more neutral and general term of "sign system".

Unfortunately, this has often just moved the debate to deciding at which degree of complexity can a sign system be called a language without a shift in meaning (such as "computer language").

I believe there is an important point to underline: a sign system does not need to be particularly complex to be interesting semiotically. Contrary to popular, tenacious belief, nature does not evolve towards more complexity, it looks for efficiency first and foremost. This implies that what remains between generations is the system giving the best possible survival-to-minimal-effort ratio. Our own species is not exempt from this rule: for example, we are more than capable, when it is imperatively needed, to use a very advanced, specific type of vocabulary; on the other hand, people will generally not know anything about the scientific jargon, technical terms or medical vocabulary if they are not involved in these circles, and it does not cause any issue in their daily life. This does not mean that they are incapable of learning these terms, of course, but simply that they do not feel like they need them, and as a result, do not learn them. Eco (1999 p. 97) pointed out an ironic facet of this idea of complexity that we have:

> Kant fait très certainement partie de ceux qui ont persuadé les philosophes de la légitimité qu'il peut y avoir à construire une phrase qui compte vingt-deux lignes dans l'édition de l'Académie sans un seul point.
>
> Translation:
>
> Kant is certainly one of these people that managed to convince philosophers that it is legitimate to build a sentence that is twenty-two lines long in the Academy edition without using a single dot.

This bit of humour is also a reminder for us to keep a cool head. A sign system will often be adapted to its emitting species and its environment. The absence in a system of a certain aspect that is present in ours should lead us to ask "why would it be there? What purpose would it serve?"

Therefore, a sign system is composed of a cognitive aspect (from nature) and a social aspect (from nurture). It functions inside a group and as a result corresponds to the part of or definition of all languages being linked to a specific community. Its complexity comes from the processes and interactions it requires (such as learning processes) and produces (such as interactions between individuals understanding the same code) rather than from its form.

When presented as such, it seems like nobody has any problem with this view of language. The debate really is just about definitions, with a large influence from ideologies: if the definition of language is such a sensitive topic, it is not because it is particularly complex or calls upon units that are difficult to evaluate, but rather because of the idea that language should remain exclusive to humans. The definitions evolved as more discoveries on the subject came out and the debate never really stopped. Frans de Waal perfectly illustrates this process. When he and his

Fig. 2.1 Drawing of the experiment on the mutual help of chimpanzees – the version from De Waal and his team, without the device

team were testing the capacity of chimpanzees to help each other, they had the idea to reuse an experiment from another team in which the monkeys had failed. They noticed the experiment involved a very complicated device and simply removed it from the experiment (see Fig. 2.1), and observed that the chimpanzees did help each other. In spite of this, their results received a lukewarm response; now, to satisfy the sceptics, chimpanzees helping each other was no longer enough – they had to help each other 100% of the time (which was not necessarily the case). He sums up that attitude by explaining that "as soon as we are talking about animals, the goalposts [the conditions to possess an ability] magically move on their own" (De Waal 2013, p. 170).

2.1.2 *Consciousness*

Consciousness, more specifically here, self-awareness is a very old debate, even when discussing what we know about human consciousness. Much like language, this notion was first considered to be exclusive to our species (at least in our historical and cultural tradition) before naturalists tried to associate it with other animal species, usually through experiments.

In ethology, one of the more telling examples is the mirror test (Gallup 1977), created to test the existence of self-awareness in animals. It consists of adding an odourless spot of colour on the forehead of the animal without it noticing, placing it in front of a mirror, and observing its reaction. If great apes, dolphins, elephants and even *Corvidae* easily pass this test, every single cat and dog fail it – that is to say they fail to produce a reaction that would undoubtedly confirm they identify themselves in their reflection. It is however still possible that they do have that reaction throughout their lives (which would explain the difference in reactions in kittens, that become anxious when seeing their reflection, and adult cats, who are indifferent to it) and that the absence of a reaction is the normal behaviour when faced with a phenomenon they judge harmless. But what to think of this result that was made for animals having sight as their dominant sense, like humans, while cats and dogs primarily use smell?

How can we confirm that these animals do not have self-awareness because they do not recognise themselves in a mirror, when we know their sense of smell takes priority over sight in these species[1]? Sight is not the "sense of self-awareness", nobody would question that blind human children gain consciousness at around the same age as the other children. In spite of this, we still struggle to find tests that call upon other major senses, and as a result, many species are deprived of tests that are relevant to their biology. The current consensus remains that consciousness is rare amongst the animal kingdom, when there are very few relevant tests.

For example, despite being lauded as very intelligent by humans, cetaceans have suffered from lack of experiments due to their very different nature (studies on these species started much later than studies on great apes or birds, for instance), as well as a lack of methodology. During a long period of time, it was accepted that it was impossible to confirm the existence of consciousness in this infra-order because none of the species in the whale family had ever been able to undergo the mirror test. Considering the blue whale is usually around 30 meters long and weights around 170 tonnes, it is easy to see what we mean by "non-ecological test".

This species, more than any other, helped us highlight the deficiencies and limits of cognitive tests and the necessity to adapt them to the species that are undergoing them. The discoveries of neuroscience on the degree of gyrification of their neocortex which goes beyond all known species, humans included, did contribute to the current acknowledgment of their great cognitive capabilities. Adding to this case linked to a material and physical difficulty to evaluate, there are other cases that show the difficulty to account for the specificity of species we try to test.

The case of Patterson (1986), the first researcher to have made a gorilla successfully pass the mirror test. It was a primordial test for a species so close to ours, and until then all test subjects had failed. If we accept as a premise some modesty, as in the fact that we cannot create a new ability in an animal but at best modify or upgrade an existing one, we need to ask what happened in that test. Gorillas are social animals with a relatively strict hierarchy, but since their muscular strength can be gigantic and easily lethal (some dominant males reach 300 kg), they try to avoid conflict as much as possible (Grassé 1955). As a result, since two males look-

[1] It is worth noting that kittens are scared when looking into a mirror, while adult cats seem like they could not care less and act like they could not see anything in there anymore, which is a point that has rarely been studied. Similarly, cats can closely watch a computer or television screen and react (differently depending on the cat) to other cats they see on screen.

ing each other in the eye is a sign of provocation, gorillas will instinctively avoid staring at someone's face, making the mirror test a difficult endeavour in their case. However, Koko has been used to Patterson since she was young and to the objects and strange things the researcher brings her, and knows she is not in any danger in these situations. In her mental world, her trust in Patterson is strong enough for her to confront her reflection (and pass the test). Self-awareness exists in all gorillas, but the test to demonstrate its existence is only relevant for one of them.

The cases of great apes handling kittens, wounded birds or children that had fallen into their enclosure is very telling of another type of consciousness, often left out of the debates: the awareness of another, differentiated from the self, specifically another who is "fragile".

The debate around consciousness is once again telling, not of our current knowledge on the topic, but rather of what we try or wish to look for, or find, in the species we submit to those tests. First and foremost, we look for a consciousness that resembles ours, based on the same senses, that can be evaluated in the same way, and which is, before anything else, a self-awareness – consciousness of others seeming much less interesting despite entailing complex cognitive and semiotic complexes. To sum up, it seems the debate about animal consciousness is not actually about knowing what it could be, but rather to check whether it is similar to ours.

2.1.3 Emotions

The stimulus-response coupling has long been the only type of response studied in animal biology, and a few behaviourists actually considered for a time that it was the only type of response that should ever be taken into account, including in humans (De Waal 2013).

That was a first problem in my thesis work; it is indeed difficult in some cases to distinguish what comes from reflexes or instinct, what results from emotions or reasoning. This problem notably arose in the study of the mother-child link: when does the mother act through instinct, or through affection, is our species also subjected to instinct when caring for our progeny, and if not, why would it be the only exception?

A second problem linked to the stimulus-response coupling came from this sensitive question – the coupling had been privileged over any other as an explanation to behaviour at the detriment of all other possibilities. Behaviourism considered that everything in animals was instinctual and that it was pointless to wonder about cognitive processes, mental worlds or motivations.

This assertion is once again more and more an ideological stance rather than a scientific truth. As a result, in my opinion, instead of being the proof of objectivity that endeavour was meant to be, it constitutes a strong experimental bias from the researcher even before any protocol or scientific model was set up. This shows that the debate on emotion and its study in animals is more a debate of beliefs rather than methods.

The topic of emotion in the animal kingdom has been discussed for a long time; the first to have actively studied the question was no other than Charles Darwin (1872, Tort 1996) himself. Since then, many proposals on the topic have followed one another, some notably choosing to study the way emotions work based on what we know about humans and then expanding upon that knowledge. Just as this study by Olga Klimecki and Tania Singer (2013 pp. 534–535) where they discuss the subject of emotional life through the scope of whether occurrences of empathy exist in an individual:

> Empathy is commonly defined as the human capacity to understand and share another person's emotion without confusing it with one's own emotional state.
>
> [The different facets of empathy related phenomena] range from rather automatic a primitive reactions, such as mimicry and emotional contagion (which can be thought of as precedents of empathy), to states like compassion or empathic distress that follow from empathy and are, themselves, important determinants of behavior.

In the last few years, neurosciences have been an enormous help for this field. The discovery and understanding – albeit only partially – of the way mirror neurons work has caused great strides in the study of emotions, both human and animal, and in many cases have defended the idea of a great similarity in a number of species – notably, the works of Pier Ferrari, Gino Coudé, Leonardo Forgassi (2009) and Vittorio Gallese (2009) open new research perspectives in this field which are absolutely enthralling.

Emotional manifestations are particularly interesting for a semiotics study because emotions are naturally unstable phenomena. Their expression is not always controlled, or controllable, and the room they take in an individual's mental world can become so large they take over all aspects of their life, sometimes to a pathological level. While these are extreme and rare cases, they do indicate the presence of more moderate cases in certain species that we would not think of at first.

However, the study of emotions in animals is still facing virulent accusations of anthropomorphism, leading researchers to navigate a veritable minefield when writing their essays to avoid the criticism that Goodall once faced when she decided to use the pronouns "he" and "she" instead of "it" to talk about chimpanzees. Claude Béata (2013 pp. 179–180) writes this analysis of an article from Jill Pruetz (2011):

> Son auteur avance avec une prudence de sioux sur le sentier de la guerre pour raconter cette anecdote étudiée avec toute la rigueur scientifique. [...] Il est remarquable que l'article, dont l'objet même est l'empathie, ne dise rien sur les émotions et les motivations des chercheurs. Le braconnage violent est qualifié d'incident, un voile pudique est jeté sur la façon dont le petit a été récupéré par l'équipe de chercheurs qui contrevient là – et je leur donne raison – à la règle de non-intervention au cours de l'observation. [...] Cet article est admirable à plus d'un titre : il induit l'émotion en restant très neutre et descriptif dans son approche, comme une métadémonstration de l'empathie.
>
> Translation:
>
> Its author treads as carefully as humanly possible to tell this anecdote studied with scientific rigour. [...] It is remarkable that an article that is actually focusing solely on empathy says nothing about the emotions and motivations of the researchers. The violent poaching is described as an incident, a veil of modesty is thrown upon the manner in which the young

> was rescued by the research team, who went against – and rightfully so, in my opinion – the rule of non-intervention during observation [...] This article is admirable for more than one reason: it induces the existence of emotion while remaining extremely neutral and descriptive in its approach, like a meta-demonstration of empathy.

This behaviour is not gratuitous; it is the result of a long academic, cultural and religious tradition, which maintain an almost schizophrenic relationship with emotions and animals in which animals always seem to come out on the losing side.

One time, the noble emotions, like altruism, were the proof of our superiority, and animals were presented without emotions, another time emotions were the animal part of us, like anger, and we were superior because of our reason, which animals could not have. Today, we still have some researchers trying to find out what is the "propre de l'homme": once, it was the empathy, then the ability to anticipate, then tool building etc.

Once again the controversy seems impossible to quell, not because of a lack of sufficient amounts of studies or data but rather a lack of definition that is universally agreed upon.

2.2 The Perspective of Humanities

It is understandable that the loudest voices in these debates and controversies are those of "hard sciences", as the subjects fall under the umbrella of biology, ethology or neurosciences. But the perspective of humanities is often strongly represented, which can be more surprising.

Beginning with the name – one might think that humanities have very little to do with animal studies. But as I mentioned when discussing the debates on language, the line between human and non-human is thin and some concepts that were previously considered proper to Man are now the subjects of rough debates between those who want to keep that line and those who want to get rid of it.

From these debates also came some concerns that are more relevant to human sciences, in this case, linguistics or philosophy.

The issue of defining what an animal is – and consequently what a human is – has long been the prerogative of human sciences rather than natural sciences. As those two progressed, the answers given by the human sciences have repeatedly needed to be reworked. What comes to mind is the famous reaction from Louis Leakey to a telegram from Jane Goodall (1998) in which she told about the discovery of tool handling in the tribe of chimpanzees she was studying in the forest of Gombe, in what is now the Democratic Republic of the Congo:

> « Il va falloir redéfinir l'outil, redéfinir l'Homme ou bien admettre que le chimpanzé est un Homme. »

> Translation:
> "Now we must redefine 'tool', redefine 'man', or accept chimpanzees as humans."

This eternal question of trying to distinguish between man and animal crystallised the debate, in human sciences, around language, as previously discussed.

But beyond the scientific aspect of this debate exists an academic aspect of the question, and the legitimacy of this book depends upon its answer: what should the language sciences study? Is there an animal semiotics, a syntax, phonetics? I obviously think so, and the very fact that the terms "biosemiotics" and "zoosemiotics" exist tends to support this opinion. But it is not an opinion that all agree with in human sciences, and it seemed important to take into account this point of view and show the controversy it creates.

Finally, I would like to come back to a national particularity, which is far from a general rule, but did influence my methodology and explains some of our choices. French academic tradition is somewhat particular when it comes to removing the barriers between fields, as well as the place of human sciences and the study of "noble" subjects (conscience, language, morals, emotions, memory, etc.) in animals.

When I will bring up the definition problems in the next chapter, we will see that many come from cultural idiosyncrasies that are not necessarily visible on the first glance but can turn out to be strongly problematic nonetheless. From this premise, I thought important to introduce the cultural origins of this work, as certain aspects of those can explain some issues, biases or premises that are mine and not automatically from researchers that come from different cultures, languages or academic traditions.

2.2.1 What Is an Animal?

The question of human-animal distinction is an old question that used to fall under humanities and philosophy (and in some cases theology) before it ended up with naturalists and then biologists.

While there have been several attempts to categorise animals to study them and learn about them, it is clear that animal studies were not all that interesting for human sciences; if animals were studied, it was mostly to oppose them to humans and better define our species.

On this topic, what better to quote than the excellent work on the history of philosophical discourses by Silver Rattasepp (2018 p. 9). His work is based upon the premise that the philosophical discourse on animals is never a reflection on the animal itself but more often than not a way to highlight humans. If we take this into consideration, there are very few actual reflections on the definition of "animal":

> The general claim of this thesis is that this recurrent discourse on animals in philosophy and elsewhere can be encapsulated in the following six theses:

1. The uniqueness of human nature can be determined by studying humans alone;
2. The most important characteristic of animals is that they lack something human;

3. When humans are compared to animals, humans are described as unique; when animals are compared between one another, they are described as merely differing from one another;
4. The distinction between humans and animals is something mental, or reducible to the mental;
5. Animals are discussed in general singular, as one indistinct, undifferentiated category;
6. Humans reach their true nature only when they expel or remove the animal from within themselves.

For Rattasepp (2018, p. 10), the problem is to find a definition of "animal", a method to characterise them, while most of the discourse on the topic only aimed to determine characteristics that are specific to humans – and flattering ones, if possible.

> Second and more importantly, the issue is rather that the items in the list above [Language, speech, symbolic forms, rationality, reason, intentionality, freedom, culture, history, consciousness, subjectivity, personhood, individuality, the immortal soul, self-reflection, laughter, ethics, politics, law, rights, property, experience of death, tool-making, labour, wearing clothes, feeling naked, making a fire, lying, pretending, bartering, shame, guilt, the as-such or the Lichtung of the metaphysicians…] — which can be extended indefinitely — are usually not at all derived from empirical research but by negation, by denying in nonhumans that which was first decided to be uniquely human […].
>
> Furthermore, the traits or criteria for determining human uniqueness are mental or reducible to the mental: something akin to smallpox (which is uniquely human since it has no other disease vectors than another human) would not do as the indicator of human uniqueness.

Griffin (1977, p. 29) also pointed out this ideological trend, especially the problems created by researchers that follow it (we have previously quoted Griffin in Chap. 1, Sect. 1.3.3).

I do not pretend to be able to end such a sensitive debate here. In my study, I use by default the term "animal" (and its variations) as a synonym of "non-human animal", which is to say, anything that is not included in our species. If my methodology aims to be a semiotic help to the study of those species, it obviously this help cannot be provided in the same way, at the same degree, for all species. Rattasepp (2018, p. 14) also explains that semiotics, especially Peircean semiotics, seems mostly indifferent towards the topic of human/non-human distinction.

> Reference to human uniqueness by way of symbol-use and the corresponding semiotic or symbolic threshold continues to hold sway in semiotic literature (for an overview of the status of semiotic thresholds in current debates, see [Rodríguez] Higuera, Kull 2017). Since the more prevalent Peircean conception of symbols is not easily amenable to constituting such a radical discontinuity (this would entail reconceptualising habit as uniquely human—a difficult proposition indeed), Carlo Brentari refers the genesis of this conception back to the German philosophical anthropology of the mid-20th century (Brentari 2018).

The debate in human sciences of knowing what could or could not separate humans from animals has taken so much room and energy that we feel it is important to remind that a single study model for all animals is impossible to create as species vary in behaviour, biology and semiosis.

If the methodological tools I will present are hopefully useful and worthy of interest for animal studies, it is precisely because semiotics allow a reading of emitted signs, an interpretation of semiosis, and adapts to the semiosis which varies from species to species.

It seemed important to me to come back to the debate on the definition of animal (in opposition to human) in humanities because one of the characteristics most used to try and keep an airtight distinction is language. It remains as such to this day (for example, I recently refused to be credited in a book for which my scientific expertise had been requested; the chapter on animal communication was scientifically flawed, only discussing project Nim and presenting Chomsky's language theories as the consensus in the scientific community. When faced with a refusal from the author to correct their text, I had to force the editor not to associate my name with that book). That is a problematic concept for a study like ours, as semiotics come from linguistics – which is to say, language sciences.

This raises the following question; are language sciences exclusively about humans or can there be some legitimacy to a study on animals?

2.2.2 What Are Language Sciences For?

While linguistics is not the only discipline of language sciences, it certainly is its leading one, and its opinion regarding the animal question is decisive in order to understand the way all sciences on the topic operate.

As other disciplines, primarily primatology, were now making a lot of progress and starting to bring in the necessary elements to begin a deeper study, linguistics, which had already suffered from Benveniste's terse conclusion on the subject, found itself both unwilling and unable of the ability to discuss the question of animal language. Humanities are mostly to blame for this development, in the sense that they long tried to distance themselves from animals. In the French academic tradition, studies on animal linguistics tend to get the cold shoulder – I will return to this.

On the other hand, the groundswell about animal language in various sciences which aims to bring them closer (both to each other and the Human and Social Sciences, or HSS) and cooperate, and create an ever-larger number of transdisciplinary topics does not seem to receive the same type of welcome in the HSS at all. While some projects go in this direction, the HSS appear very reluctant to the changes in methodology and functioning that interdisciplinary cooperation entails, and the logical conclusion is that articles, publications and research following that direction are relatively rare and sometimes looked down upon.

As briefly discussed in previous parts, much like any other scientific field, ethology has gone through various schools of thought and scientific ideologies. It makes sense to notice that during their evolution, some of those welcomed zoosemiotics with open arms, and others with great scepticism.

The behaviourist movement was not interested in the field whatsoever, which is not that much of a surprise, since it did not simply belong in their outline: there was no real semiosis in the animal kingdom, no use of symbols or interpretation processes and, generally, nothing that could not be explained by the stimulus-response process (Lehner 1996).

An opposite reaction came from other biologists and ethologists, mostly – but not exclusively – primatologists, who quickly understood the potential that such a field could have on their research, and did not hesitate to quickly start using linguistic terms, like the Gardners using the word "conversation" while discussing chimpanzees (Bodamer and Gardner 2002).

The idea of a biological approach of semiotics, or a semiotic approach of biology, will progressively gain ground until becoming a scientific field involving all species, including ours (Sebeok and Umiker-Sebeok 1992).

After Thomas Sebeok, who is seen as the father of the field as it is today, zoosemiotics found itself a *de facto* orphan, deprived of a real leader, and relegated to a minor branch of semiotics, which had its own difficulties finding its spot amongst the language sciences and linguistics.

However, after his passing, the texts that had helped create the field, or the works he had inspired began to spread – at least in some academic circles, this was not the case in France for example, but I will get back to that. It is notably the case for Peirce's, which remain barely available in the French-speaking academic world specifically, and the Latin world in general, Italy being an exception. I briefly discussed what Peirce's semiotic theories allowed us to do in animal semiosis studies. From there, it is not difficult to understand that the theories and positions from biosemioticians and founders of zoosemiotics were picked up by Peirce's continuators, mainly Eco. From then on, zoosemiotics benefited from the support of renowned linguists to cement its legitimacy as its own disciplinary branch:

> *Zoosémiotique : elle représente la limite inférieure de la sémiotique car elle étudie les comportements relatifs à la communication à l'intérieur des communautés non-humaines [...]. On pourra d'autre part vérifier si au niveau animal existent des comportements qui, dans une certaine mesure, peuvent être définis comme « culturels » et « sociaux ». L'aire sémantique de ces termes s'en trouve élargie et, par conséquent, notre notion de culture et de* société (Eco 1984 p. 14).

Translation:

Zoosemiotics: it represents the lower limit of semiotics because it studies the behaviours relating to communication in non-human communities [...]. We could also see if some behaviours exist on the animal level that, in some way, could be defined as "cultural" and "social". The semantic area of these terms is broadened as a result, and consequently, so would our notion of culture and society.

2.2.3 *The Specific French Academic Tradition*

Zoosemiotics is a rare discipline in the landscape of French-speaking research in general, and French research in particular. Semiotics itself really is not all that much represented, compared to other fields under the umbrella of linguistics and language sciences. The teachings and studies that exist are almost exclusively based on the works of French-speaking semioticians, mainly Ferdinand de Saussure et Algirdas Greimas. Peirce, my main source to build this methodology, is almost completely missing from the picture – mostly because he has not been translated very much. The choice to undertake a doctorate thesis in this field was thus done after a long decision-making process, as the field suffers from a lack of easily accessible secondary literature as well as a lack of academic consideration. Indeed, until recently, the animal was not deemed a worthy subject in linguistics to build a career around. The choice of a doctorate thesis in zoosemiotics was a challenge, and this finds its roots in the history of linguistics and language sciences in France (this note can also be valid for many others Latin countries).

The Latin word "anima" was the word used to talk about the life, the vital principle, but soon was used to talk about the soul (in French, "anima" became "âme", the word for soul). Then, there was a separation between the animal "the life being" and the human and its "âme", "the soul being".

The idea was widely used by philosopher René Descartes (1637) and his follower Malebranche (1674). To talk about animals, he created the concept of "animal machine". The meaning was that animals were like clocks: moving, making noises, with delicate pieces and even a beat, but without soul, ideas, language or emotions.

This idea was very encouraged by the global occidental culture he (and we) lived in: a monotheist culture, with Man, made in the image of God, above every other creature. This led to a conflict between reason and emotions, in which the animal seemed to always be losing. One time, the noble emotions, like altruism, were the proof of our superiority, and animals were presented without emotions, another time emotions were the animal part of us, like anger, and we were superior because of our reason, which animals could not have.

We have seen just before that one of the major definitions for this kind of work, the definition of language, was directly correlated with a lot of controversies, linguistical, philosophical and ethical ones. A question must be asked as this point: why should we need a definition of language for this work? The point is that, in the French academic traditions, as in others too, there was this believe that important or complex things could be expressed only by the way of language. We don't agree with this position, and will tell more in the definitions' section. But as an example, we consider emotions as the failure of the language, the moment it becomes totally unable to properly express what we are living – and then, producing a lot of signs during this conflict, allowing semiotics methodology to work. But we had to explain what language was first to then defend this position. The point is that, a bit like for emotions, nobody can today completely agree on a language's definition, because most working on it try to create a definition that would completely exclude the animal, but this is today impossible without excluding some humans too.

The French academic tradition seems very attached to its definitions. This characteristic explains why the controversies and debates in its midst are more virulent than they are in other academic traditions or scientific circles, even if those do have their own. It also explains why I chose to make definitions of one the main focal points of methodology.

This tradition also orientated our thought process and positions, it is an integral part of the genesis of our writing this methodology. The omnipresent controversies and debates about animal studies in humanities led us to think about the origins of those debates, the need to take them into account and to take a stance in them. The specific problem of definition has led to its own thought process and treatment, which will be the topic of the next chapter.

2.3 Author Position

After highlighting the main controversies on the topic, it is possible to answer the question that was asked at the beginning of the chapter: what are the controversial aspects of the animal issue from a semiotics point of view that must be account for in this type of work?

If the various controversial aspects have been described and made explicit, I have not yet discussed how to take them into account. I would like to do so now before ending the chapter. I will answer our question in two steps.

First, I would like to give and justify my position for the writing of this book and my methodological work in general on the entirety of the problematic aspects I discussed. I hope this will allow us to clearly state the reasons that make us choose this or that position on a certain topic, how this choice influences the way my methodology is built and what possible biases could come from this choice. I stand by the opinion that any choice, even a philosophical one, from the researcher can be a source of bias in their work; by making my own clear, I hope to reduce biases to a minimum, or, if I cannot completely erase them, to examine them from all sides in a more efficient manner.

Therefore, even if it is obviously not perfect, I consider that my position answers, on every aspect, to the logic of the "lesser evil" in the choices it has to make.

Some points however need to be addressed in a second step. Many of the previously mentioned controversies had a definition as their starting point, consequence or tension point. This assessment led me to a certain position when it comes to dealing with definitions, their use and their assimilation into our methodology.

Before making this the focal point of our next chapter, I would like to clarify and justify this position, so as to make clear the reasons why I will come back to certain definitions – sometimes at length – that seem obvious for everyone.

2.3.1 The "Lesser Evil" Position

In the previous debates, aside from the issue of definitions that I will be dealing with separately, a common denominator that I mentioned in previous chapter surfaces: the anthropocentrism that is not openly stated. If language, conscience or emotions are controversial subjects, it is because they are seen as concepts that need to be exclusive to humans and that attributing them to other animals can only be an anthropomorphic mistake. It is this violent rejection – which is sometimes justified, if we recall the Clever Hans story (Rosenthal 1970) – of anthropomorphism that partly explains some stances in these debates.

However, while it can be a trap that scientists should avoid, not everything about anthropomorphism should be dismissed. This phenomenon also observes certain rules, especially the fact that it applies to species that indeed are close to us, and in which it is possible that thought or cognitive process exist (Christen 2011). To take an extreme example, we tend to more easily think that great apes, as our close cousins can feel anger, sadness or jealousy than we do with mosquitos, who we are very distant from on a phylogenetic level.

"Close" and "distant" certainly do not mean "identical" which is why researchers rightfully tend to put some distance between themselves and their subjects. However, in many cases, this intention to keep a certain distance created entire experimental biases that did not exist originally. Lieberman (1977 p. 23), on the topic of language, explains that this intention comes from our phylogenetic isolation, which leads us to want to find differences in kind rather than degrees between us and other species:

> The difference between human speech and human language and the communication systems of other animals may not be qualitative. It is difficult to think of any aspect of human behaviour that is really unique. [...] the difference may be only a quantitative phenomenon. Qualitative behavioural differences can be the result of quantitative structural differences. [...] The differences between human and animal communication are more obvious because the intermediate stages of hominid evolution are no longer alive.

I will illustrate this point using two distinct examples – one intentional and one accidental – which should not be treated in the same way.

The first example is a famous one. Jane Goodall, the now-famous primatologist submitted observation reports on Gombe chimpanzees at the beginning of her career. The review panel criticised her use of "she" and "he" pronouns and demand that they are replaced by "it" (Goodall 1998). She will also receive criticism regarding her interpretations of the subjects' behaviours and emotions, but this is not related to this current analysis.

What can we say of this demand? Is it relevant to talk objectively about the study subjects? First, we know that in chimpanzees, the behaviours, roles, family bonds and life cycles of males and females are very different, so using a pronoun that species the gender does actually appear to be a token of good consideration of field observations. Back then, this demand from a scientific panel seemed to wish for some "moral etiquette" (not using the same pronouns as we do for humans) at the expense of descriptive rigour, which is not really a show of objectivity.

What is even more curious is that Goodall justifies the use of those pronouns by the great diversity of "personalities" found in the studied subjects and, in her work, defends the possibility of a chimpanzee "person". As a result, the choice she makes in her writing style is relevant and logical, and as long as the demonstration that follows is coherent and well-founded, there is no reason to consider the use of gendered pronouns a mistake. History and the progress of research in this field have largely proven her right since then.

Let us now discuss an accidental case. It is a good illustration of what Faroul Seif (2012 p. 137) calls a distortion linked to human-animal dualism:

> The dualistic separation between the realms of zoosemiotics and anthroposemiotics is a fallacy. Although the notion of "semiotic animal" is conceived to overcome the human-animal dualism, still many supporters of zoosemiotics remain anthropocentrically biased. And despite the fact that the late works of Darwin have radically changed the scientific perception and conceptualization of animal semiotics, the dualism remains distorted toward human idiosyncrasy.

Our example is a meta-analysis gathering a series of experiments on both monkeys and humans for comparison (Boesch 2007). It goes without saying that conditions need to be as similar as possible for experiment results to be comparable. The researchers made sure to put forward the same test, in the same lapse of time, with the same equipment, to monkeys (young and old) then humans (young and old). The results found a clear superiority for humans.

However, one point was neglected. A flagrant difference in treatment was found between human subjects and monkey subjects – presence, or lack thereof of experimenters, one-way glass, parents for the young – without any of this ever being accounted for while interpreting the results. It is not a case of deliberate ill-will or negligence from the researchers, but simply rather of a lack of the same consideration towards monkeys as towards humans. And this lack of consideration – one could almost call it a lack of anthropomorphism – quickly led to a series of anomalies in the method.

What are the consequences? First, the researchers wanted the humans to feel at ease and unafraid by the experiment, which is normal. So they stayed with the subjects, who were, more often than not, in a familiar room with the added experiment equipment. Inversely, they estimated that the monkeys would not have the same worries, and we highlighted earlier that most experiments (16 out of 23) were made with subjects in a test room, separated from the researchers by a one-way glass.

Moreover, the human child could stay in the arms or on the knees of a parent as much as the experiment allowed it. If that was not possible, a parent or familiar adult would stay in the room, within sight. There is no need to bring up that the young monkeys did not benefit from similar treatment. We can easily understand what happened with our current knowledge of the species. A youngling without its mother feels at best greatly unsafe, at worst in grave danger (Despret 2009). At that moment, its mental world is filled with a priority much superior to solving the experiment in front of it. Odds are the same difference in success rate would have been obtained between a human child tested with the conditions listed alone, and a child left alone in an unknown room.

In order to confirm this hypothesis, it should be reminded that some of the experiments were done in similar conditions for both species, although it was a minority. Those are the ones to have resulted in similar performances in monkeys and humans.

It was indeed the lack of "consideration" for the tested animal that was at the origin of this bias. However, other biases can exist – in the experiment protocol, in the conclusions drawn from the results, in the experimental devices – that simply come from researchers wrongly understanding the way a species operates.

The anthropocentrism that is so strongly defended in those debates can turn out to be a methodological disaster. For all of these reasons, I believe that when choosing methods, the benefit of the doubt should go to anthropomorphism rather than anthropocentrism when the experiment data did not allow a clear decision. This stance has consequences, particularly when it comes to the ethical management of results, which I will return to in Chap. 7.

2.3.2 *About the Particular Case of Definitions*

Peircean semiotics offers a theoretical framework, a work method and research leads, but the author's works always stayed rather unburdened from the constraints of terminology and disciplinary restrictions (Peircean semiotics has been found to have uses in speech analysis just as in cryptology). This does not remove scientific accuracy, on the contrary, the demonstrations based on this theoretical framework must always be well-argued and consistent; but this method allows one to avoid falling into the pit trap of too rigid a model. However, I precisely aim to avoid putting "narrow limitations" on a mostly unknown field. I mostly lean on the thoughts on the topic of someone who also was a fervent "populariser" of Peirce, Jakobson, (1973 p. 11):

> En réalité, la plupart des désaccords récents sont dus en partie à des écarts de terminologie et en partie à une répartition différente des problèmes linguistiques choisis et signalés par les chercheurs […]. Il arrive en fait que pareille sélection aboutisse à confiner la recherche dans des limites étroites et à faire négliger les sujets qui ont été écartés.
>
> Translation:
>
> In reality, most recent disagreements are partly due to differences in terminology and partly due to a different allocation of linguistic problems chosen and highlighted by researchers […]. Such selection can sometimes confine research to narrow limitations and neglected subjects that were pushed aside.

The problem is particularly well illustrated in the case of defining emotions, which is why we will be using this example to explain our position.

I am far from the first linguist to work with emotions and thus far from the first to face the various problems they can bring, beginning with a semantic and lexical perspective.

Algirdas Greimas (1981) explained this quite well when he discussed anger. He showed that under the term "anger" – that we seem to be able to easily agree upon –

existed a large palette of lexemes of different intensities and length. The examples of rage, rancour or bitterness illustrate his point.

However, it is doubtful that these examples win unanimity in zoosemiotics and semiotics. If an individual appears to show bitterness towards another, it seems difficult for systematically consider that a form of "anger".

Rage seems to be less of a question. And even then – in the examples I studied and analysed during my study, anger regularly appears to be an extreme tell of an emotion that is not necessarily anger. It should not be forgotten that anger is, for example, an integral part of the sorrow caused by death as well as the grieving process. Therefore, should the semiotician classify that rage as a sign of anger or sorrow?

The same question can be asked about jealousy, which sometimes leads to spectacular outbursts of rage. This is actually a twofold issue. First, should these outbursts be considered as a sign of anger, or pain, caused by jealousy? Second, it should be understood that considering jealousy as pain is a cultural aspect – in some cultures, any love is possessive love, and as a result jealousy is not a sign of pain, but rather a sign of affection.

Here we are reaching the first of many challenges in a research work in zoosemiotics, whether it is a work on emotions or any other complex concept. It is a challenge that does not necessarily wish to be solved but rather to question our lexicology and actively think about it.

It is indeed unthinkable – and this perspective is consistent with the detailed choice in the next chapter about the definition of emotion – to solve a problem that is far from just being semantics, but rather mostly cultural. At the beginning of my study, I wrote my propositions while being fully aware that the categories I was referring to as "anger", "jealousy", "sorrow", "joy" etc. were cultural categories that most likely were unconscious choices. After a while, before using this or that term, I took the time to learn what it meant on a chemical and metabolic level (Bohler 2007), at least for our species, to objectify the various "frontiers" between the different emotions as much as possible. These terms remain relatively arbitrary nonetheless.

But it does not make them erroneous. I simply believe this point should be reflected upon, especially since a recurring argument against animal emotions studies (or language, or self-awareness, or the transmission of cultural know-how) is to say "this is not really jealousy" or "this looks like sorrow, but it is not" or even "one might say it is distress, but it is not the same kind of distress as the one we feel". This point of view has already been expressed by Greimas (1988, p. 343):

> « Les définitions ne sont que des expansions des dénominations, et, de ce fait, substituables les unes aux autres [...]. »
>
> Translation:
> "Definitions are nothing but the expansion of denominations and as a result are exchangeable with each other […]."

This argument is all the weaker if we look at it from the perspective we just brought up, which is to say the nature of what "really" is jealousy, sorrow or distress is an impossible question to answer in human cultures. These terms often end up mixing and overlapping and moving from one culture to the next and blurring the lines between each emotion.

Those moving frontiers allow us to bring up a new reflection point: is it really relevant to categorise emotions? From a real, material point of view, an emotion can be described – grossly – as a chemical mix, the appearance of this or that molecule, in a certain density, added to another or blocked by this or that antagonist telling our brain how it feels, whether it is normal or pathological. The various treatments against depression in humans actually do similar things: they try to fix the chemical balance of the brain to stop the "parasite" emotions.

At which point does the level of a substance or the proportion of the mix tip over just enough for us to be able to say "we are no longer able to say it is emotion X" or "we can observe a feeling of Y"? The question can sound absurd. It is probable that emotion categories raise the same issues as their animal counterparts: they are fixed categories in a moving, continuous living element.

This is not to say we should no longer use these terms. First, to return to an earlier metaphor, we should behave like the physicists who keep in mind that the quantic model does not describe what is real, but something close enough to reality to be studied. Second, and I will return to this, these terms remain useful because they speak to the way we feel, and that is very real.

My opinion on definitions is most likely very particular. I believe that they hold no more meaning than what we are willing to give them, since they are above all tools that answer a need to scientifically classify and categorise. As a result, it is important to sort definitions, establish the ones we need during a work, a study, a determined methodology, and the ones we do not need, and could also be an added source of controversy with no added value.

References

Agha, A. (1997). "Concepts" and "communication" in evolutionary terms. *Semiotica, 116*, 189.

Armony, J., & Vuilleumier, P. (Eds.). (2013). *The Cambridge handbook of human affective neuroscience*. New York: Cambridge University Press.

Baldwin, J. M. (Ed.). (1901). *Dictionary of philosophy and psychology*. London: Macmillan.

Barbieri, M. (Ed.). (2006). *Introduction to biosemiotics: The new biological synthesis*. Dordrecht: Springer.

Béata, C. (Ed.). (2009). *Zoopsychiatrie : L'attachement*. Marseille: Solal Editeurs.

Béata, C. (2013). *Au risque d'aimer*. Paris: Odile Jacob.

Benveniste, E. (1966). *Problèmes de linguistique générale*. Paris: Gallimard.

Bodamer, M., & Gardner, A. (2002). How cross-fostered chimpanzees (*Pan troglodytes*) initiate and maintain conversations. *Journal of Comparative Psychology, 116*(1), 12–26.

Boesch, C. (2007). What makes us human (*Homo sapiens*)? The challenge of cognitive cross-species comparison. *Journal of Comparative Psychology, 121*, 227–240.

Bohler, S. (2007). *La Chimie de nos émotions*. Genève: Aubanel.

Brentari, C. (2018). From the hiatus model to the diffuse discontinuities: A turning point in human-animal studies. *Biosemiotics, 11*, 331–345. https://doi.org/10.1007/s12304-018-9329-8.
Christen, Y. (2011). *L'Animal est-il une personne ?* Barcelone: Flammarion.
Darwin, C. (1872). *The expression of the emotions in man and animals*. Londres: John Murray.
De Waal, F. (2013). *Le Bonobo, Dieu et nous*. Lonrai: Les Liens qui Libèrent.
Descartes, R. (1637). Discours de la Méthode.
Despret, V. (2009). Il est temps d'en finir avec Harlow. In C. Béata (Ed.), *Zoopsychiatrie : L'attachement* (p. 13). Marseille: Solal Editeurs.
Eco, U. (1984). *La structure absente: introduction à la recherche sémiotique*. Paris: Mercure de France.
Eco, U. (1999). *Kant et l'ornithorynque*. Paris: Grasset.
Favareau, D. (2006). The evolutionary history of biosemiotics. In M. Barbieri (Ed.), *Introduction to biosemiotics: The new biological synthesis* (p. 5). Dordrecht: Springer.
Ferrari, P., Coudé, G., Gallese, V., & Forgassi, L. (2009). Having access to others' mind through gaze: The role ontogenic and learning-processes in gaze-following behavior of macaques. In C. Keysers & L. Fadiga (Eds.), *The mirror neuron system* (pp. 239–249). Hove: Psychology Press.
Gallese, V. (2009). Mirror neurons and the social nature of language: The neural exploitation hypothesis. In C. Keysers & L. Fadiga (Eds.), *The mirror neuron system* (pp. 317–333). Hove: Psychology Press.
Gallup, G. G. (1977). Self-recognition in primates: A comparative approach to the bidirectional properties of consciousness. *American Psychologist, 32*, 329–338.
Goodall, J. (1998). Learning from the chimpanzees: A message humans can understand. *Science, 282*, 2184–2185.
Grassé, P.-P. (Ed.). (1955). *Traité de zoologie, anatomie, systématique, biologie, Tome XVII. Mammifères, les ordres : anatomie, éthologie, systématique. 1er fascicule*. Paris: Masson.
Greimas, A. (1981). De la colère : étude de sémantique lexicale. *Actes sémiotiques – Documents, 27*.
Greimas, A. (1988). De la nostalgie : étude de sémantique lexicale. In *Hommage à Bernard Pottier vol. 1 (p. 343)* (Vol. 7, pp. 343–349). Paris: Klincksieck.
Griffin, D. (1977). In T. Sebeok (Ed.), *How animals communicate* (p. 29). Bloomington: Indiana University Press.
Guillaume, A. (2014a). Animal : « être sensible » unanimement désensibilisé. Sémiotique du sensible. *Revue trimestrielle de la Fondation Droit Animal, Éthique et Sciences, 81*, 35–37.
Guillaume, A. (2014b). L'interthéoricité : sémiotique de la transférogenèse. Plasticité, élasticité, hybridité des théories. *Revue PLASTIR, Plasticités, Sciences et Arts, 37*, 1–36.
Jakobson, R. (1973). *Essais de linguistique générale, Tome 2 Rapports internes et externes du langage*. Paris: Éditions de Minuit.
Keysers, C., & Fadiga, L. (Eds.). (2009). *The mirror neuron system*. Hove: Psychology Press.
Klimecki, O., & Singer, T. (2013). Empathy from the perspective of social neuroscience. In J. Armony & P. Vuilleumier (Eds.), *The Cambridge handbook of human affective neuroscience* (p. 534). New York: Cambridge University Press.
Lehner, P. N. (1996). *Handbook of ethological methods*. Cambridge: Cambridge University Press.
Lieberman, P. (1977). In T. Sebeok (Ed.), *How animals communicate* (p. 23). Bloomington: Indiana University Press.
Malebranche, N. (1674). De la recherche de la vérité. Où l'on traite de la Nature de l'Esprit de l'homme, & de l'usage qu'il en doit faire pour éviter l'erreur dans les Sciences.
Neveu, F. (2004). *Dictionnaire des Sciences du Langage*. Herstal: Armand Colin.
Patterson, F. (1986). *The mind of gorilla: Conversation and conservation. Primates: The road to self-sustaining populations*. New York: Springer.
Peirce, C. S. (1901). Sign. In J. M. Baldwin (Ed.), *Dictionary of philosophy and psychology* (Vol. II, p. 527). London: Macmillan.
Peirce, C. S. (1992). *Ecrits sur le signe*. Paris: Éditions du Seuil.

Pruetz, J. D. (2011). Targeted helping by a wild adolescent male chimpanzee (Pan troglodytes verus): Evidence for empathy? *Journal of Ethology, 29*, 365–368.
Rattasepp, S. (2018). *The human mirror. A critique of the philosophical discourse on animals from the position of multispecies semiotics*. Tartu: University of Tartu Press.
Rattasepp, S., & Bennett, T. (Eds.). (2012). *Gatherings in biosemiotics*. Tartu: University of Tartu Press.
Rodríguez Higuera, C. J., & Kull, K. (2017). The biosemiotic glossary project: The semiotic threshold. *Biosemiotics, 10*, 109–126.
Rosenthal, R. (1970). Le préjugé du maître et l'apprentissage de l'élève. *Revue française de pédagogie, 13*, 39.
Sebeok, T. (Ed.). (1977). *How animals communicate*. Bloomington: Indiana University Press.
Sebeok, T., & Umiker-Sebeok, D. J. (Eds.). (1992). *Biosemiotics: The semiotic web 1991*. Berlin: Mouton de Gruyter.
Seif, F. (2012). Semiotic animal in a transmodern world: Hovering between zoosemiotics and anthroposemiotics. In S. Rattasepp & T. Bennett (Eds.), *Gatherings in biosemiotics* (p. 137). Tartu: University of Tartu Press.
Tort, P. (Ed.). (1996). *Dictionnaire du darwinisme et de l'évolution. Vol.1*. Paris: Presses Universitaires de France.

Chapter 3
Necessary and Problematic Definitions

Abstract Conception sub-question n°2: what definitions should we all agree upon from an interdisciplinary standpoint?

First of all, this chapter will introduce the definitions considered necessary for a semiotic work about animals. The suggested definitions will revisit the concepts of emotion (in its general sense), consciousness and memory. Later on, it will present the cases deemed problematic regarding definitions: the concepts of language, emotions (in a more specific sense: fear, grief, anger etc), intelligence and culture. Solutions will be submitted each time to avoid the issue of definitions.

Keywords Animal studies · Definitions · Lexicology · Language · Culture · Emotions · Intelligence · Memory

While I know that the definitions used in this type of work are the source of a number of debates, they remain a necessary part of research work nonetheless.

Classifications and categories are powerful tools, and in semiotics, which deals with sign categories, combination and interpretation, it is necessary to have definitions to work with.

In this chapter, I will answer the second conception sub-question I brought up in my introduction: What definitions should we all agree upon from an interdisciplinary standpoint?

To answer this, I will split this chapter into two categories: one that deals with necessary definitions in a semiotics work on animals, and another dealing with definitions that raise more problems than they help solve.

In the first category, I will discuss the definitions of emotion, conscience and memory, concepts I deem essential for our methodology model to function properly.

I will dissect these concepts, offer definitions, explain why I chose or built those definitions rather than others, what their uses are in a semiotics work on animals, but also what their limits are and even the biases that might come with them and that we should watch out for.

In the second category, I will deal with the definitions, or rather the difficulties of defining the concepts of language, intelligence, culture and specific emotions. While

P. Delahaye, *A Semiotic Methodology for Animal Studies*, Biosemiotics 19,
https://doi.org/10.1007/978-3-030-28813-6_3

I think the general concept of emotion is necessary for this type of work, I will also show why the definitions of emotions and categories of emotions (fear, sorrow, anger, jealousy) are complex and problematic.

Concerning language, intelligence and culture, these are very interesting concepts that I ran into regularly during my research, but they prove to be so difficult to define, so broad or reductive that they cause more problems for those trying to work with them than they provide substantial help as conceptual tools.

I will explain in detail why that is, and offer solutions in each case to try and deal with the problem; without denying it, but also without making it a central point of the methodology.

3.1 Necessary Definitions

It is obvious that it is neither possible nor advisable to work without a single definition, whether it is a work in life sciences or language sciences. Therefore, despite the controversies they may create, certain concepts need clear definitions that are universally agreed upon. This agreement is assumed as being interdisciplinary, which also means that in semiotics research on animals, it must encompass the vision of the two affected disciplines with the same precision and rigour.

The first concept that I thought necessary to define is emotion. It was at the centre of our research work and is the direct cause of this methodological model.

From a semiotics point of view, it is particularly interesting as it is a phenomenon that creates a large number of signs in an individual experiencing it, signs that, for a semiotician, are entry points to the inner workings of the subject. Therefore, it was necessary for this concept to have a clear and accurate definition that would serve as a solid basis for future interpretations of those signs.

The second concept I wanted to take time to define is conscience. As we have seen in the previous chapter, it is one of the most controversial aspects in animal studies.

But since I will talk about the categories of sign emissions (voluntary, involuntary but conscious, unconscious) during the next chapter, this concept must be thoroughly thought out.

This will be all the more true afterwards, when will we discuss the ethical challenges of a study such as ours, especially in dealing with "foreign consciences" where the concept of conscience will, logically, be at the centre of the problem, with major implications.

The last concept I thought necessary to define is memory. It may seem anecdotic but it has the distinctive feature of not having a similar meaning at all in language sciences and life sciences. A work in zoosemiotics must therefore clarify this potential source of misunderstanding.

This is one of the concepts that can have major implications in ethics for animals (see Chap. 7), and with great reach comes great responsibility.

3.1.1 Emotion

Emotion is a subjective phenomenon that is difficult to grasp. Neurosciences, psychology or even endocrinology can shed some light on how it appears and operates but it is difficult to precisely describe it. One of the most complete characterisations in this sense that I found is by Aaron Ben-ze'ev (2010, p.61):

> An emotion is described as being generated by perceived significant changes; its focus of concern is personal and comparative; its major characteristics are instability, great intensity, partiality, and brief duration; and its basic components are cognition, evaluation, motivation, and feelings [...]. An emotion is part of a greater affective realm, which also includes sentiment, moods, affective traits, and affective disorders [...]. This mode is the most complex, comprehensive and dynamic of all other mental modes.

This description gives us all the elements to not only understand what an emotion is but also how to study it properly. It gives us its trigger (what generates it), its characteristics (what it is made of), its focus (what it targets), its context (what place it holds in the life of an individual and what it entails in their life), its components (what mental abilities and processes must be present for it to exist) and its status (its level of complexity and adaptability).

This attempt at a definition,[1] extremely thorough and almost exhaustive (in its original version – here we chose to shorten it a bit for the sake of clarity) is valid for all emotions, human or animal, simple and complex, in the scope of what they all entail. It allows a precise reading of various emotional manifestations, and an opportunity to visualise what may be at that moment in the "black box", to borrow a phrase from Umberto Eco (1999).

But since it encompasses all emotional manifestations, it does not allow us to, amongst other things, differentiate between a simple emotion and a complex emotion. The latter is what interests us most, both for the complexity of the cognitive or social processes involved and the breadth of signs it causes. However, it will give us criteria (particularly for complexity and involved processes) with which to build a definition of complex emotion that allows us to distinguish it from its "simple" counterpart.

It is indeed through the definition of emotional complexity that we will be able to separate different emotions. While the criteria of this complexity are not universally agreed upon in the scientific community at the moment, everyone agrees that it does exist; the issue comes with knowing where to place that distinction and which criteria to use as a basis.

The characterisation of emotion done by Ben-ze'ev offers different leads to find these criteria, some of which were borrowed from researchers in various fields. For example, the fact that an emotion can have such a degree of complexity and importance that it creates affective disorders (something a simple emotion could not do).

[1] In this case, it is an attempt at defining emotion in the general sense, the problem of various, specific emotions will be discussed in the next part and will create more encounters with the issue of subjectivity.

Certain manifestations of stress or sorrow are in this category, and present themselves so often in the animal kingdom that some neuropsychiatrists now study these human disorders through animal models (Sufka et al. 2009).

Others have directed their attention on the focus point of emotions. While it is first and foremost centred on the self (implied in the definition by the use of "personal"), a complex emotion in an individual generates a comparative recognition of emotion in others and a manifestation of empathy in those around them. It is recognised and understood. Watanabe and Kuczak (2013, p. 180) attest to such a manifestation of emotional recognition:

> Similarly, a series of elegant studies has shown that chimpanzees understand the emotional content of their species-typical facial expressions of emotion (see review by Parr and Maestripieri 2003).

For some other researchers, an emotion is considered complex in the sense that it makes the life of the individual that experiences it more complex, by adding new levels of "mental life" such as the systems of value (Zahn et al. 2013) or the feeling of morality (Prinz 2010). For others such as Greimas (Greimas and Fontanille 1991, p. 93), it is the idea of length and stability that will create this distinction, even if he maintains that, in this domain, all attempt at a definition is cultural first and foremost:

> On peut montrer aisément, à cet égard, que [...] le fondement du système reste relatif à une culture donnée.
>
> Pour le **sentiment**, on retiendra qu'il est donné comme un état affectif complexe, stable et durable, lié à des représentations.
>
> L'**émotion**, quant à elle, serait une réaction affective, en général intense, se manifestant par divers troubles, surtout d'ordre neuro-végétatif.
>
> Translation:
>
> We can easily show that, in this regard [...] the foundation of this system remains relative to a specific culture.
>
> For the *feeling*, we should remember that it is defined as a complex, stable and durable affective state linked to representations.
>
> The *emotion* would be an affective reaction, generally intense, manifesting itself through various disorders that are mostly neuro-vegetative.

Nonetheless, for this study I will prefer a definition that focuses more on the characteristics of complex emotions and how they function, allowing us to categorise manifestations more easily with the smallest possible margin of error:

> Les émotions secondaires sont plus complexes et concernent des centres cérébraux plus haut placés dans le cortex. Elles peuvent comprendre les émotions fondamentales de peur ou de colère, ou des émotions plus nuancées, comme le regret, l'envie ou la jalousie. Les émotions secondaires ne sont pas automatiques : elles sont traitées par le cerveau, l'individu y réfléchit. (Bekoff 2009 pp. 39-40)
>
> Translation:
>
> Secondary emotions are more complex emotions, and they involve core emotions of fear and anger, or they could be more nuanced, involving such things as regret, longing, or jealousy. Secondary emotions are not automatic: they are processed in the brain, and the indi-

vidual thinks about them and considers what to do about them – what action is the best one to perform in a certain situation. (Bekoff 2007 p. 8)

This aspect appears primordial to me. The individual does not experience that emotion and then forgets about it a few seconds afterwards. They are about to think about it, transform it – for example, someone gets scared by something, and after the initial jolt, they become curious and try to understand what it was – or at least try to transform it (a sad of stressed individual looks for others to help or comfort them). This way of thinking about emotion and the place it takes in someone's mental world will generate visible and analysable manifestations that will be the focal point of our study.

It goes without saying that not all these manifestations are done through language – far from it – hence the importance of semiotics in my study. This is not necessarily a problem when it comes to emotional experience:

> We have seen that the individuating aspect of a thing can offer purchase (albeit sometimes uncertain purchase) to those uses of language that break their frame of semantic convention (comprising their defining inferential commitments) to show the content of experience rather than say it. (Pugmire 2010 p. 384)

As I mentioned earlier, emotions are one of the rare aspects in one's mental life (including in our species) where language is often unable to let someone properly transmit to others what they need to "say".

3.1.2 Consciousness

Consciousness in animals has already been the subject of experimentations, the most famous one being the mirror test. I have already discussed how this test could have its limits. As a result, it would be ill-advised to reduce the definition of consciousness to just the ability to recognise oneself in a mirror, even if passing this test is an undeniable proof of consciousness. But what about those that do not pass this test? Is it caused by a lack of ability, or a lack of relevance from the test?

Until recently, the test was very cut-and-dried: the individual/species either passed the test, or failed it. But De Waal and his team (De Waal and Robert 2002, pp. 118–119) suggest the consideration of an entire range of nuances. De Waal says:

> Aziut, un macaque à longue queue de l'Institut de recherche sur l'identité en Inde, joue tout le temps avec les miroirs. Il a appris spontanément à les utiliser pour voir ce que les hommes et les chiens font derrière son dos. [...] Aziut joue aussi avec le reflet mouvant de sa main dans un miroir [...] et il essaye parfois de maintenir deux miroirs l'un en face de l'autre en en maintenant un au sol avec son pied et l'autre au-dessus de sa tête. Ce ne sont pas exactement les jeux d'un animal qui serait trompé par les miroirs.
>
> Permettez-moi d'ajouter l'exemple des singes rhésus femelles adultes [...]. À chaque saison des naissances, nous voyons les femelles placer leur nouveau-né sur le sol, faire quelques pas et regarder attentivement vers l'un des miroirs [...]. Elles semblent relier leur propre comportement (placer leur enfant au sol) à l'image dans le miroir. Le fait qu'elles n'utilisent pas le miroir, comme les chimpanzés, pour regarder leur propre reflet, peut rele-

ver du peu d'intérêt que les singes portent à eux-mêmes comparé à l'intérêt suscité par des créatures aussi attirantes que leurs nouveau-nés.

Translation:

Aziut, a long-tailed macaque from the Identity Research Institute in India, is always playing with the mirrors. He spontaneously learned to use them to see what people and dogs are doing behind his back. [...] Aziut also often plays with the moving reflection of his hand in a mirror [...] and sometimes tries to have two mirrors facing each other by keeping one on the floor with his foot and another above his head. These are not exactly the games that an animal fooled by mirrors would play.

Allow me to add the example of adult female rhesus macaques [...]. Each birth season, we see females place their new-born on the ground, take a few steps and attentively look at one of the mirrors [...]. They seem to connect their own behaviour (placing their child on the ground) to the image on the mirror. The fact that they do not use the mirror like chimpanzees to look at their own reflection may be due to their lack of interest towards themselves, compared to the spike of interest caused by fascinating creatures like their new-born.

This kind of observations can only incite us to avoid cut-and-dried interpretations and better take into account the possible nuances that exist. Because it is logical to think they are numerous, as one ability, when faced with the same incident, can express itself completely differently based on the biology of the species.

These observations, and the arguments discussed earlier, lead me to propose two points of reflection that relate to the handling of nuances in zoosemiotics in general, and in the case of evaluating consciousness in particular.

The first point is that I strongly think there needs to be a lot of caution regarding binary tests. Not due to the way they function – I have already shown what they have taught us – but rather what we project on them as scientists and researchers.

Language has long been seen as an impassable frontier separating humans from the other animals, a frontier that, if we could manage to cross it, would allow us to gain access to their thoughts, feelings, consciousness and view of the world. But nothing is less certain.

Ludwig Wittgenstein, who I have already quoted, is doubtlessly one of the most clear-sighted language philosophers when it comes to animal semiotics. Other than the fact that his works shed a light on the reasons behind the successes and failures of tests using intermediary sign system, Wittgenstein also foresaw their limits. When he wrote "a lion could speak, and we could not understand it" (Wittgenstein 1961, p. 356), it was not about denying the possession of language to other species, but in all likelihood about pointing out that our difficulties in understanding animals do not come from language itself but rather of a biological and cognitive whole that differentiates every species from each other and makes them fundamentally foreign to one another. A universal definition of conscience, amongst such diverse species, seems to be quite unrealistic.

The second point of reflection is that, as obvious as it seems to us, the concept of self-awareness is probably not adapted to the whole animal kingdom. While we can wonder if that ability exists in certain species, it may simply be irrelevant for other species.

Throughout this book, I have used and will continue to use the terms "subject", "individual", and even – although much more rarely – "specimen". While these terms can sometimes cover slightly different realities (for example, I will consider that a subject is a being acting in the context of semiosis, which is not necessarily the case of an individual, who can just be a witness or even be there by happenstance and have no particular reaction), they are not problematic themselves. I did not find ideological positions from researchers showing a need to use a certain term above another for conceptual or methodological reasons, and, while there has been some debating about considering an animal as a "subject" rather than a simple "object of study", I have made my decision from the start of the study by resting upon reliable specialised sources, and I believe it is not necessary to return to this topic.

However, a debate is possible (and advisable) on another point: knowing whether these terms represent the same realities across different species. Our custom of having pets has led us to consider each physical "cat" or "dog" individual as a "human" physical entity: a living biological body with physical characteristics, and requiring certain physiological parameters to be constant (regular nutrition, sufficient hydration, fixed body temperature, etc.) without which the individual will die. By extension, each physical individual is also a mental individual that may have self-awareness: we know that cats, just like humans, need to eat, but we are not particularly surprised when we notice that one cat is a glutton, another is a picky eater, and another is very fond of a food that is actually toxic; just like we are not surprised that some of our fellow humans cannot stand the bitterness of coffee, refuse certain types of food because of ethical or religious beliefs or drink to excess despite the known harmful effects of alcohol. Each physical individual has its mental counterpart.

But is it relevant to say that these two "individuals" coexist in all species? This point is where the question of collective intelligence is interesting and opens a debate, particularly in light of the most recent discoveries on the species that possess it.

For example, some advanced calculating abilities in ants were discovered (Dorigo and Stützle 2004). Calculating itineraries, infallible internal functioning, the ants regularly stun entomologists and defy the calculating power of our computers (we will get back to this aspect in the "Intelligence" part of this chapter). Yet what are we really saying when we report "ants can calculate"? All ants? All physical individuals in ants? Only in groups? How many ants are required before a "calculating" entity can exist? All of these are particularly tricky questions.

But the fact is, the physical individuals, on their own, are very much devoid of an ability that their group possesses. To remain logical, we need to create two different levels of individuals amongst these species – and other similar ones, mainly bees, and to a lesser extent, termites:

- Physical individuals: each ant is a living biological entity that can be identified by its physical characteristics and requires certain conditions to be respected so as to stay alive.

- The cognitive individual: each ant is incapable of elaborate cognitive activity, but the anthill behaves as an individual capable of calculations or complex organisations.

Beyond the simple fact of being able to explain why it would be pointless to try and find some form of cognition or emotion in an isolated individual in this type of species (which is an interesting but specialised fact), this distinction also allows us to shed some light on another fact. The concept of an individual, and thus of self-awareness, has a relevance that is much more limited than we tend to think. While I chose the example of collective intelligence to illustrate this point, other species can make us raise the same question (Chapouthier et al. 1980).

Self-awareness is a complex act linked to the individual's mental world and the way they feel. However, not all physical individuals constitute a cognitive individual. It is not pointless to search for consciousness in this or that species because it shows very little intelligence, but because the physical individual is not necessarily the cognitive individual in that species. At the same time, it must be understood that this complex, extremely subjective act can have modes of expression that strongly vary from species to species. If we define self-awareness as the ability to consider oneself a complete individual, different from one's peers and the world, then we need to pay attention to all the signs that can reveal a logic of the self, and not just the ability to recognise one's reflection. For example – I will bring it up later on – the tests on the feeling of injustice in monkeys can offer some proof of this type of semiosis: being angry due to being a victim of injustice entails the ability to distinguish between the self and the others, between what belongs to others and what belongs to oneself and to understand that the wronged party is "me".

3.1.3 Memory

Memory seems like it has little to do with semiotics, but it actually is a major notion once code (and consequently communication) becomes the topic. Without memory, neither transmission nor progress is possible because the individuals always have to start over from scratch. This reason was brought up recently to explain why some species, such as many cephalopods with apparently very intelligent individuals, make little to no cognitive progress: their mode of reproduction entails that parents rarely, if ever, survive the birth of their young. As a result, generational transmission is impossible, and the species cannot move towards a more structured thought, a more evolved intelligence, and certainly not an organised culture.

For the theorists Dan Sperber and Deirdre Wilson (1989), long-term memory is a foundation of reasoning that is in perpetual motion. It is not rigid; that is what allows cultural, linguistic or technical progress in a group. If that memory was too fixed and refused all re-assessment of knowledge and assets, progress would simply be impossible. In humans, there is a risk that cultural tradition or acquired knowledge forbids putting things into question and closes the door to progress. Simply enough, this phenomenon can also be found from one generation to the next; it is possible to think that as brain plasticity diminishes, it becomes too difficult for long-term memory to put acquired knowledge into question or update information.

In the species studied by the observer, the presence of a long-term memory is easy to demonstrate: can the individual remember someone that has died a long time ago? Can they tell memories involving them? Do they learn from an experience, even long after it happened? Do they show a semiosis of the memory?

A being possessing long-term memory, even if that is all they have, can be considered a being with specific memories, their own identity, a capacity for emotion linked to memory. They must be treated as a being capable of feeling, and as such those feelings should be accounted for and respected.

Another aspect of learning that Sperber and Wilson gave a lot of importance to is the ability to use a coding-decoding system, that is to say a system that allows transmitting information in such a way that it can be understood by all members of the group. If this ability comes naturally to many species, it is not the case for all of them: suricates (Manser et al. 2001) and songbirds (Konishi 1965; Marler and Mundinger 1971) are some of the many species that need to learn the code used to communicate with their peers.

The necessary mastery of that system raises a question: how does the individual learn it? This question could be a troubling *mise-en-abyme:* how do you learn the tool required to learn everything? Of course, we do not pretend to have a definitive answer to a divisive subject. The interest we have in it is the observation of the group's semiosis, of the parents or "educator" of the youngling that is still devoid of that system, and the specific behaviour that notes an understanding of the fact that that individual is still "mute" to the group (use of a pseudo-language, cognitive stimulation or, on the contrary, abandonment due to the young's incompetence).

Another interesting, yet completely different point to observe is the ability that many species, including humans, have to learn snippets of other species' coding-decoding system. Through observation (of the individuals interacting with each other, or of the signs and behaviours with a causal link, like a cat's yowling before it attacks), individuals manage to absorb a part of the other species' system and initiate an attempt at language. These exchanges are extremely curious, as sometimes two individuals without a language in common manage to understand each other. These cases can particularly be found between humans and their pets: cats and dogs do not speak any human language but understand congratulations or reprimands, they are sometimes capable of extremely evolved suitable responses and can tell very subtle emotional variations in the voice or behaviour of their owner (sorrow, disease…). Similarly, while no owner can properly reproduce any sound of an animal language, many are able to tell the difference between, for example, the type of meowing expressing a need for food and the one expressing unease or discomfort, maybe due to a disease.

As a result, memory takes on a particular definition; it is not just the remembering of events, but also an ability to create a connection between sign and meaning, and through that connection, to interpret any code. While it seems secondary, it turns out to be a key concept for the study of communication, emotions and cognitive processes of an individual.

The best way to be convinced of its importance is to observe what happens when memory is lacking. The learning process is the first mechanism to be compromised by memory's breakdown. It is a complex and delicate mechanism that needs to rely on a lot of prior knowledge and automated processes to function. If those two functionalities happen to fail, nothing new can be absorbed by memory.

Memory defects can compromise the learning process from their first manifestations. First, because new information is often – for reasons we have yet to understand – the most fragile when dealing with the weak growth of memory. The subject that shows such symptoms is doomed to repeat the same learning processes over and over again. Little by little, if the issue causing that cognitive loss continues to grow, data or mechanisms predating the issue can end up disappearing as well, making any new attempt at learning even more difficult: due to the lack of tools, it is even more delicate for the new information to settle. It can become harder (attempt to compensate for the missing tools) or longer (attempt to recreate those tools). This is why the first signs of Alzheimer cerebral degenerations in humans is the lack of ability to learn new things, disorientation due to immediate events not binding properly and short-term memory loss.

If simple defects can have such an impact on a human brain, they can have heavy consequences on an animal brain. Firstly, because it will not have had the opportunity to create compensation mechanisms that exist in humans to help the failing memory for a time. And secondly because if memory is "blocking" any new learning process, it will stop an individual, who might be remarkably developed on an intellectual level, from acquiring knowledge, techniques, cognitive reflexes and more evolved intellectual abilities.

If the memory defect worsens, it will impact other abilities. Therefore, a new question regarding information transmission comes up. The more damaged the memory, the harder it is to maintain a long, well-built reasoning.

Therefore, in the case of animal studies, it is possible to think an individual can perfectly make technical discoveries or elaborate simple concepts but find themselves unable to efficiently transmit these ideas. It is also possible that their interlocutor, who is dealing with the same cognitive issues, is incapable of accurately memorising the information that is being transmitted – as best as the individual can – and the problem intensifies if this interlocutor now tries to transmit that information to yet another individual in the group.

Thus it is easy to understand where the issue is in an animal group – since every member in the group suffers from a seriously faulty memory, it is not possible to have any long-term transmission. There might not even be any possible simple transmission either, which is even more crippling; how can they agree on a code, remember the objects linked to the signs, remember how to organise thought to transmit it? The individuals in the group may be very intelligent and capable of understanding complex concepts on their own, but the transmission is so impaired by the serious faults of the memory that those ideas remain cloistered in their mind: to each other, they are "mute".

At its most advanced stage, the loss of memory (or its absence) results inexorably in the loss of identity, which is built from memories, experiences, emotions felt for a loved one. All these elements are tightly knit to memory, and if it disappears, so do they. This does not necessarily mean that self-awareness is impossible without it, because it is not an identity, it is just the conscience of the existence of one's own thought.

Sufferers from the terminal stage of Alzheimer's most likely still have self-awareness, but their identity has completely faded away. This notion is often echoed by their family members and loved ones: the person in front of them is not the one they have known and loved. That person is already dead – psychologists that monitor those families call this "ambiguous loss" – and all that is left is a body, an empty shell that is unable to connect with them in any way.

In animals, the consequences are obviously different. This whole or partial absence greatly cripples intellectual evolution. By preventing an individual from building their own identity and from being conscious of it, the heavily flawed memory blocks complex conceptualisation for the individual.

In humans, it also appears that some terminal Alzheimer's patients no longer react to any exterior stimulation. As memory disappears, constructed thought becomes harder and harder: it is impossible to remember what was being thought about just seconds earlier and relate it to something else. If something attracts them, they can spend hours staring at it, with the same deconstructed, brief thoughts on repeat.

In animal species, the total absence of memory prevents all complex thought. In the total absence of constructed thought, not only is the intellect relatively poorly developed, but there is also good reason to believe that the tools required for those abilities to appear are also absent, or else the abilities would have developed.

There may be an exception in collective intelligences: since they do not need individual consciousness because individuals do not matter, or thoughts proper to a specific member of the group since the group thinks as a "network", maybe such a species can do without individual memory and only possess a collective one that somewhat escapes our understanding. The model I am presenting now is not adapted to these species, at least not yet, and the definition of memory given earlier does not apply to them.

3.2 Problematic Definitions

The sub-question I wanted to answer in this chapter, that is to say "what definitions should we all agree upon from an interdisciplinary viewpoint?" is faced with a problem. All the concepts that are interesting to use in a semiotics study on animals are concepts that are impossible to find interdisciplinary agreements about. In this second part, I will describe the four most problematic amongst them.

First and foremost is the notion of language, which I have largely discussed in the previous chapter on debates. A controversial topic for linguists, this concept also suffers from heavy ideological tension. While it is difficult to propose a definition that everyone will agree on, it is possible to study *why* it is not possible to have everyone agree, and maybe propose another term that better fits semiotics (and not linguistics) and allows for research that is less charged by opinions.

Secondly, as I briefly discussed in the introduction to this chapter, while it is necessary to propose a definition of emotion in the general meaning of the term, it is very tricky to provide a precise and definitive terminology for each different emotion. Many of them overlap, interfere with each other, are named and seen differently depending on the culture and appear or disappearing based on the sensitivity of each individual. Moreover, it seems particularly complex to try and provide objective, definitive definitions to phenomena that exist in the subjective view of an individual.

The third concept is closely linked to conscience, which I defined earlier – this concept is intelligence. Whenever we try and study the animal and its semiosis, its internal functioning, its cognition, the concept of intelligence always ends up involved in the analysis. But it is a vast, hazy and controversial concept, even when we talk about our own species. Worse still – it is tempting, during an animal study, to compare the abilities of the subject with those of humans, with the same risk of an anthropocentric bias consisting in viewing our species as the best, only allowing the animal to equal us at most. All of this makes the concept of intelligence a difficult one to grasp, and any attempt at a strict definition leaves itself wide open to a spiral into anthropocentrism.

And finally I will discuss the concept of culture, a concept that has made many appearances in my work while analysing semiosis of transmission, learning or rituals (of reconciliation, mourning, etc.). The word "culture" is a very sensitive term, and the tension surrounding it is similar to the one you will find with language. Not only has culture long been seen as a human prerogative (we were beings of nature *and* culture, while other animals were just of nature), but also, the attempts at defining it inside our own species has sometimes led to philosophical, ideological and political drifts.

3.2.1 Language

The distinction between the definitions of language and communication has been a controversial topic amongst linguists, particularly when it comes to animals. Most books dealing with animal communication give it an extremely broad definition that is much too general to allow it to be called "language":

> The most common and most widely accepted definition is that animal communication is behaviour involving signals as information carriers. Signals in animal communication are understood as behaviours or traits that evolved specifically to provide information from a sender to one or more receivers. (Naguib 2006 p. 277)

This is very much a general definition of communication, which could just as well be applied to vegetal communication. As a result, it seemed sensible and coherent for us to keep calling this "communication".

But the animal kingdom also has a wealth of systems that are much more complex and demanding on a cognitive level. Specifically, three aspects held our attention as linguistic and cognitive elements that could qualify a mode of communication as "language": the precision in the form of the message, the voluntary aspect of the act of communicating and the complexity of the content.

It did not seem useless or intellectually dishonest to deconstruct and rebuild that definition, if the rebuilt definition is closer to the reality of the facts and leads to more rigor. Here we lean on the words of Greimas (1988, p.343): "Definitions are nothing but the expansion of denominations and as a result are exchangeable with each other […]".

To build a definition of language that takes these aspects into account and allows a distinction in the animal kingdom between what is communication and what is actual language, we began with the definition from the *Dictionnaire de la linguistique* (Mounin 2004 p. 196)

Langage : L'aptitude observée chez tous les hommes à communiquer au moyen des langues.
Langue : Tout système de signes vocaux doublement articulés, propre à une communauté humaine donnée.

Translation:

Language: The aptitude observed in all humans to communicate through tongues.
Tongue: Any vocal sign system with double articulation that is proper to a specific human community.

From there, it is necessary to work step by step. First, I removed from this definition what appeared to be unjustified anthropocentrism. As a result, "language" becomes "the aptitude to communicate through tongues" and "tongue" becomes "Any vocal sign system with double articulation that is proper to a specific community". It seemed essential to keep the notion of a specific community because it is not limited to our species while also revealing an interesting point about the way language functions – the social and communal aspect.

Then, I focused on what seemed too reductive, even regarding our own species. Is there no human tongue that is non-vocal? Of course there is – nobody would say nowadays that sign languages are not languages, that deaf and mute people are deprived of language. Do these languages have double articulation? It is tempting to answer "no", since the very notion of phonemes is irrelevant in a non-vocal tongue. It is true that it is possible to spell a proper noun in sign languages by using a sign for each letter of the alphabet. However, this system is more of a strict borrowing connecting the sign language and the written language of the same community (as a result, French Sign Language has 36 signs in its dactylological alphabet to reflect the 36 signs of the Latin alphabet, while Japanese Sign Language, for which the very notion of alphabet is a problem, has 45 signs and four diacritics to spell the written language). It seemed tricky to talk of double articulation in those languages on the basis of such an outlying phenomenon. So, "tongue" becomes "any sign system that is proper to a specific community".

To avoid juggling between terms, I apposed the two definitions to create a single one for "language":

Language: "The aptitude to communicate using any sign system proper to a specific community.

This definition is actually relatively close to the one given by de Saussure (2002), that is to say, the general faculty of expression through signs, the main difference being that he believed it was only applicable to human communities.

The final step was adding the important points that seemed to have been missing from the previous definitions: the fact that we are talking about voluntary communication when we are talking about language, and the complexity that the term "language" entails, both in form and content. Without adding these aspects, our definition would fall in exactly the same trap as the first definition we reviewed – perfectly adequate for communication, but far too general to fit the name of "language".

Therefore, I could propose the following definition:

"Language: the aptitude to voluntarily communicate using any complex sign system proper to a specific community."

However, this reconstruction has a major issue. We have seen how the debates around language are structured once the question of animals comes up, and how they were charged with the ideological aspect of the subject. Even a reconstructed definition of language cannot get rid of this aspect. My proposed definition does not escape that rule, and I think the problem must be contemplated differently, in this case by taking a term that has not been the object of so many debates, and use it as a new tool where the denomination of "language" has become too dangerous a tool to use. To do so, I have chosen the term of "dire", translated by "saying" both because it indicates the more general act of semiotic production and the very interesting fact that semiotic production is a consequence of the existence of something "to say".

It is obviously a strong linguistic term that could be the subject of its own thesis. I do not intend to give a universal and definitive definition of the act of "saying", but rather a possible definition, another perspective on this semiotic act.

The reason for this is that the notion of "saying" gives language the conditions to be as I defined it earlier. Because there is something to "say", because if this thing is important, it explains the efforts in complexity and diversity that appear in language performances. Of course, the "saying" does not answer the question "why is there language?" (if there even is an answer to that question), but it does explain "why this form", "why this content" and "why this semiotic performance in this exact moment in these circumstances".

"Saying" is first and foremost formulating a precise message, no matter its form, its code or its mode of transmission. Works in primatology will, once again, open new research paths for us in a clear way. For example, Zuberbühler (2013, p. 415) presenting works on the language of Diana monkeys leads to unquestionable conclusions: the cry used to signal a bird of prey is not the same as the one alerting others of a danger on the ground, and they are understood and treated differently by the receivers of the message (remaining still or going down to the ground for aerial attacks, scattering or going towards the individual responsible for the alert for a threat on the ground). They are also different from the ones used by a dominant male to indicate the presence of food, launch a group movement or communicate with the nearby females (those cries are also treated differently from each other).

But it is even more subtle than simply that; the dominant (as well as some other adult males) has a method of specific cries (a combination of two distinct cries, which can also be used independently) which increases the degree of urgency of the alert it replaces. Zuberbühler compares this to the imperative.

Suricates, also known for their vocalisations, have around 30 distinct alarm cries, each related to very specific cases, which is an even more advanced performance (Manser et al. 2001).

With this first point, one could argue that Diana monkeys and suricates fulfil the first criterion of the notion of "saying" which is one member of the group emitting messages with a clear goal (alerting about a certain type of predator, with a certain degree of urgency, and expecting a certain reaction from the group) and the rest correctly interpreting such a message (their understanding is made explicitly clear by the absence of a surprised reaction from them when faced with the predator they had been warned about). Tests have also proven that when the warning was "rigged" (the scientists would show a fake snake to the alpha, who would employ the "ground" warning cry, then they would broadcast the screech of a crowned eagle, a predator announced through a different alert), the reactions from the monkeys were actually loud and intense, as the males would relay the new threat and the other members of the group would radically change their behaviour.

The particular term of "precision" I use about the emitted messages comes from the observation that the same sign (signifier) is always treated by the receivers as having the same meaning (signified). In an added degree of precision, a combination of simple signs can acquire a new meaning that is different from the original and individual meanings of each sign.

"Saying" is also having the wish to communicate and a reason to do it, or at least try to do it. What Koshik (Stoeger et al. 2012) does by imitating the words of his carer is most likely not language, but it is however a behaviour that allows him to "say" something here: his strong wish to communicate with an individual of another species whose language he learned to mimic so the carer would understand his intention. There needs to be a sizeable wish to communicate and a great need for close social ties (as a reminder, his carer spends a lot of time with him on a daily basis, but this still seems insufficient for the pachyderm) in this individual to justify the amount of complex effort put into this endeavour. This simple fact already gives us a lot of information on his emotions and his mental world, a world in which social ties are such a necessity for his well-being that it warrants that technical research and the efforts to adapt.

There is one last aspect of the notion of "saying" to study after form and justification; content, and particularly contents, which we will call "complex contents". If an individual expresses a complex emotion through any means, it indicates the existence of all the cognitive processes that are necessary for that emotion – in the case of sorrow from mourning, it entails the abilities to forge strong, long-term social ties, to consider an individual as a unique, non-interchangeable entity, to tell the different between sleep or sickness and death, and to understand the latter as a definitive, irreversible state.

Emotions are a particular aspect of the notion of "saying" due to what they reveal about the mental complexity of the individual expressing them and what they can lead them to, but also because they are, by nature, what we need to "say"; and language often is not sufficient to accomplish this. Putting mourning, passionate love, anger, or what art makes us feel into words is tricky business, and complex emotions often are the ones to almost universally render language "devoid of meaning". Jakobson even presents this insufficiency of language as the origin of the poetic function in the first tome of his *Essais de linguistique générale* (Jakobson 2003), making the function an attempt to express what is normally inexpressible through another means of language.

Emotions are not the only kind of complex content. Processes that entail referring to time over a long period, calling on a value system, being able to see someone else's perspective or invention are also extremely demanding on a cognitive level and should be considered complex contents. Concretely, a non-exhaustive list includes all that will be used to "say" morals (this point alone is the source of many questions and will be dissected as I continue this study), memories, empathy (which is closely related to emotions on one side and morals on the other), learning and lying ("semiotics is [...] the discipline studying everything which can be used in order to lie") (Eco 1979 p. 7).

Complex contents, even when not expressed through language, are an important criterion of what "saying" is in our study's framework. The subtitle of my thesis was made with this in mind: there are important, complex things to "say" that rule over whole parts of our lives, sometimes without the need or possibility to use language.

For all these reasons, I believe that the question of "saying" is more interesting and relevant in a work like ours than the question of language proper, which is more controversial and stuck in the mud of definition issues where the question of "saying" is on what is almost virgin land. A semiotic study on animals must not try to find a distinction between what is language and what is not, but study any and all semiosis that allows one to "say" something.

3.2.2 Emotions

While the previous part of this chapter showed that it was crucial to have an attempt at characterising the general concept of emotion, for lack of a true definition – it is simply not the same for specific emotions. First off, even their number is difficult to determine: it varies depending on the literature, culture and language. In Norwegian, the word "Forelsket" refers to a particular fraction of the feeling of love, the intense and euphoric part at the start of a dalliance, in Japanese, they use the word "love", pronounced [lɔvɯ], to indicate a specific type of love that is intense and expressive, similar to "Western" love, and it is not included in the "ai" of familial and marital love, etc. The vocabulary used to describe emotions can be extremely vast and subtle (see Fig. 3.1), and vary from one culture and language to the next.

A second problem is that not everyone agrees on what is or is not an emotion. The issue of art is particularly interesting on this point: the feeling of something being beautiful, the aesthetic pleasure, the wish to create – are these emotions?

Several cases of animal "artists" have been reported, but the best testimony when it comes to monkeys comes from the ethologist Desmond Morris (1962), who studied painting monkeys. He highlighted several interesting factors in the paintings produced by the monkeys he studied. First, they have a precise idea of what they want to create and do not tolerate having their canvas taken away before the painting

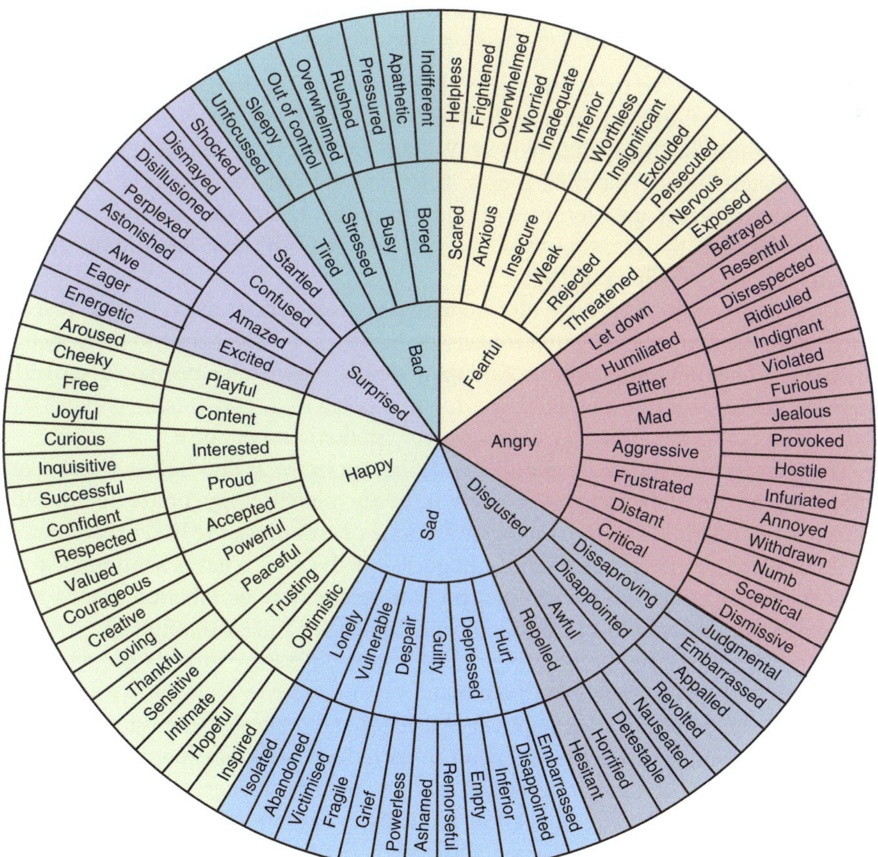

Fig. 3.1 The wheel of emotions is a widely used therapeutic tool, but the distribution of its segments and the number of spokes vary from one language to the next

is finished, indicating that they see it as a supersign that is incomplete if the painting is not finished. Second, the works most liked by art critics are systematically spontaneous ones: every attempt to make a monkey paint in exchange for a reward resulted in a lacklustre painting. Lastly, these works have their own signatures and are not easily imitated by humans. Yves Christen (2011, p. 336) writes about this in *L'animal est-il une personne?:*

> Élément intéressant : [les toiles] ne sont pas particulièrement faciles à imiter. Ceux que l'exercice a tenté n'ont produit que des peintures maladroites et dans l'excitation. [...] Il semble que les chimpanzés peintres aient aussi une idée très claire du moment où ils estiment leur œuvre terminée. Si on la leur arrache des mains avant terme, ils le prennent fort mal. En revanche, une fois achevée, elle ne semble plus guère les intéresser.
>
> Translation:
>
> Interesting point: [the paintings] are not exactly easy to copy. Those tempted by the endeavour only produced awkward artworks and in excessive excitement [. . .] It seems that painting chimpanzees also have a very clear idea of the point at which they consider their work finished. If we take it away from them before it is, they get extremely angry. On the other hand, once it is finished, they no longer seem interested in it.

Elephants also have their fair share of "painters". Two females, Siri and Carol, would draw by holding a pencil with their trunk. The former had a fervent admirer of her drawings in the Dutch painter Willem De Kooning, the latter created an impressive amount of paintings, now scattered around the world. A renowned authority in abstract art, Jerome Witkin from the University of Syracuse, had acknowledged all the artistic impact before even learning that the author was not human (Gucwa and Ehmann 1985).

Surprisingly, octopuses have also been impressive – once again – in this respect. In the wild, they pay a lot of attention and dedicate a lot of their time to the way they place shells and coloured rocks around their shelter. There have been no discoveries made so far regarding a potential use for this behaviour, and scientists have concluded that octopuses probably just enjoy the aesthetic aspect of it.

These examples showcase the difficulty in determining and define what is and what is not an emotion, and how to study it. In my methodology, I have decided that since it is not possible to agree on the definitions of emotions, the best way to handle the issue is to only keep a few large categories, defined by what is semiotically relevant for each of them. For example, pleasure can easily be labelled thanks to its external signs, and whether it is the physical pleasure linked to animals grooming each other, the pleasure of mothering or the aesthetic pleasure linked to beauty, what matters most is that we can give it a name.

Some of the emotions I kept are very basic (anger, fear, pleasure, sadness) – others are more complex, but were difficult to include in the first categories (jealousy, which could have been anger, affection, which could have been pleasure). One is a delicate subject as it is difficult to know if it truly is an emotion, but we have decided to include it, due to its vast presence and impact in the animal kingdom: empathy.

The consensus amongst biologists nowadays is that empathy comes from the existence in an individual of neurological cells called "mirror neurons". They work by generating feelings the subject has already experienced that relate to what it sees happening to another. This is well known to humans, who, when seeing someone get hurt for example, will start wincing.

Beings capable of showing empathy not only have a sufficiently developed intellect to accomplish the relatively complex task of putting themselves in someone else's place, but also possess the emotions they attribute to others (otherwise, the mirror neurons would have no effect).

Washoe is doubtlessly one of the subjects that taught us the most in this regard. Fouts (Fouts and Mills 1998) recalls the moment when a carer came back after a leave of absence due to a miscarriage, and explained to the female chimpanzee that she had been gone because her baby had died. Washoe responded by signing "cry".

At that time, Washoe had already lost her own child, Sequoyah, from a respiratory infection (see Fig. 3.2). The activation of her mirror neurons sent her a strong emotion when listening to the carer telling her story. Moreover, we know that the death of her child had been heavily emotionally charged for the chimpanzee, who had spent nights standing vigil over the sick young and long demanded his body using ASL once it had been taken away from her.

Consequently, her use of the "cry" sign towards a human telling her about the loss of her child leads us to think this sign was not chosen by coincidence nor was it used lightly, especially as monkeys cannot biologically cry out of sorrow. To her, it was the sign that represented best what she felt when her child died, and therefore, what she believed the young woman felt at that moment.

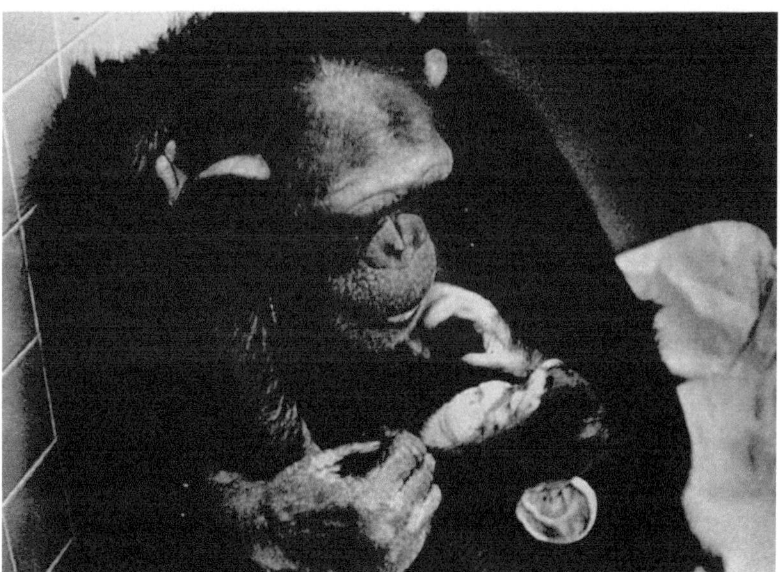

Fig. 3.2 Washoe lost her two biological children before Fouts entrusted her with Louis, her adopted son

But that is not all. At that point of the conversation, Washoe simply wanted to know why the carer had been gone for so long, but did not show any affective distress that required an immediate response from the chimpanzee. We could consider her response as "free", that is to say, she wanted to show her empathy towards the carer with no visible need in return, and did so over time, as when the carer was leaving for the night, Washoe signed "hug, please". According to Fouts, that is a sign that she could imagine the pain of the young woman and the need to be comforted over time, as Washoe had needed.

The choice to study an emotion in an animal group must be a semiotic choice first and foremost: it must be based on all the available signs that could grant us as access point to emotions and cognitive processes, and not on fixed categories that say there are this or that many emotions and we need to see if a species has the capacity for all of them. It is the observable (directly or as the consequence of emotions) signs that must show us what to study and not a pre-established list that tends to lean towards anthropocentrism. Indeed, that viewpoint works on the assumption that we humans have the "full set" of emotions – the animals being, at best, our equals.

I am not calling into question the comparative and analytic approach; I see it as interesting and productive, as previously mentioned, in evolutionary biology. This approach also exists in other aspects of biology: there often are studies trying to find out how bad the eyesight of snakes is, the maximum possible frequency that dogs can hear or whether cats have a sense of taste as developed as ours. But in these studies, the opposite questions also exist – that is, trying to understand the senses some other species have that we do not. For example, we know that the mantis shrimp (*Stomatopoda*) possesses 12 photopigments and 16 cones (Thoen et al. 2014), which doubtlessly gives it an outstanding colour vision which must make ours look quite dull. We also know bats are capable of very acute echolocation to find their way, while only a few people blind from birth are capable of, and in a very limited capacity.[2]

But in the domain of studying emotions, we have not been able to find, during our 3 years of research, the question of whether – just as with the physical senses – beings biologically different from us could feel some types of emotions we could not.

While there certainly are several explanations for this, the first is simply a cognitive issue. It is particularly difficult for us to imagine an emotion we cannot feel, similarly to the case of the mantis shrimp, where it is impossible for our brain to imagine a new colour. But regarding the shrimps, we do have measurable points of reference (the photopigments and cones) that allow us to explore that possibility. We still struggle to establish similarly precise points of reference regarding emotions within our own species – to find some amongst foreign species is a pipe dream for now. As our abilities currently stand, it is the semiosis of an emotion that shows us it exists, however we call that emotion.

[2] A fascinating compilation of methods, studies and testimonies is available on the website of Daniel Kish. http://www.worldaccessfortheblind.org/

3.2.3 *Intelligence*

When studying complex phenomena (emotions, language, culture, techniques) amongst the animal kingdom, the concept of intelligence is a recurring topic, sometimes in a comparison effort (saying an animal is as smart as a child of a certain age), and sometimes in a qualitative manner (saying that a certain behaviour is the proof of a certain category of intelligence). From this observation, I decided to look into the biases that we could have when talking about intelligence – we, as in all of us researchers, and I as in the author of this specific study.

The definition of intelligence is, in itself, problematic, as it is the case for all concepts in this part of the chapter. It is however possible to circumvent this point as we have done so far, by using a list of identifiable, debatable and measurable characteristics instead of a definition of the word that tries to be universal. Therefore, this is not an insurmountable challenge if previous studies on the aforementioned characteristics happen to exist, which – thankfully for us – they do.

There is however a real difficulty that is the perfect example of the potential cognitive bias that comes with the position of the experimenter. If we are so eager to find sign of intelligence in other species, it is because we, as humans, are particularly well-equipped in this regard (or at least, we believe this enough to have named ourselves *Homo sapiens*). But it is extremely difficult to study objectively something we have always been immersed in, even if we have not realised this and have not been studying it most of the time.

The risk encountered here is the same the first linguists to study animal language faced – on a different subject, but with the same reasoning. It is also for this reason that we wanted to bring up this topic before introducing methodological tools and processes: because the risk of this type of bias in a study on this type of subject is constant.

Coming back to the case of collective intelligence, we see that there needs to be a very specific situation (a species with great intelligence but with no cognitive individuals) that is also fairly rare in the animal kingdom (social hymenoptera are an exception even within the arthropod taxon) for the emotion-intelligence, or self-awareness-intelligence, or narrative memory-intelligence connections to be put into question. It is interesting to note that it is just this ultra-specific category that Benveniste (1966) had chosen to reopen and close the debate on animal language, forgetting that, if animal intelligence is different from ours, the different kinds of intelligence of different species sometimes are even further apart from each other. While it remains almost certain that an advanced form of cognitive functioning is required to develop emotions, self-awareness or long-term memory, the opposite is not true; just because we see a species without compassion, altruism or sorrow does not mean it is not relevant to look for signs of intelligence within it.

Considering the risks of confusion and potential biases that come with the term "intelligence", we are of the opinion that it should simply not be used in a semiotic study on animals. The value of semiotic studies is due to their work on signs, and therefore the manifestations of internal processes. If signs can be observed, then the cognitive processes exist, and it is most likely more interesting and useful to describe, explain and interpret them than it is to know if they come under the concept of "intelligence" that we cannot even agree upon to qualify our own species.

As far as I am concerned, I will not use this word in my methodological thought process, rather choosing to speak of "cognitive processes", "cognitive abilities" and "cognition", which I see as less controversial as they only indicate that some known but particular processes (triadic awareness, theory of mind, projection in time, etc.) are occurring, without presupposing to know which, and how many, are necessary to consider an individual "intelligent".

3.2.4 Culture

The notion of culture in animal studies is cause for debate. Culture has long been the element that distinguished Man, the cultural being, from other animals, natural beings. But the definition of culture is complex and we often disagree on the subject in our species. While we may lack a precise definition, I focused on several characteristics of culture in order to find an agreement and possibly use this concept in animal studies.

First, I can point to the complex operation of transmitting technical and encyclopaedic knowledge amongst a social group, both between and within a generation.

Generational transmission can be seen through particularly favoured interactions between the adults and the young of the group (as a reminder, the notion of "youth" varies: in great apes, orang-utans become adults between 7 and 9 years old, chimpanzees at 10, and bonobos and humans become sexually mature between 13 and 15 years old). These interactions can focus on different domains; they can be about encyclopaedic knowledge, social codes, moral codes, rituals or skills and know-how.

A skill is a tricky and complex element to transmit. At this time, the question of how certain skills are transmitted between generations remains unanswered – that is the case for the well-known "sandals" chimpanzees use to climb kapok trees (Nishida et al. 2010) – but some of those transmissions have been observed and what they taught us is fascinating. The nut-breaking skills for nuts from *Coula edulis* and *Panda oleosa* trees is much more complex than it appears, and the years required for a youngling to learn them (which are spent "training" under adult supervision) are very costly both in time and energy for the involved adults that know these skills, which is a visible sign that the generational transmission of skills is very important to them (Goodall 1986).

It appears straightforward that skills with such a degree of complexity cannot be learned and transmitted without a particularly acute and elaborate form of communication. Studies led on great apes undergoing experimental training seem to show these animals are completely conscious of their actions and the process to accomplish, even if it is very new to the species – we can quote the experiment reported by Whiten et al. (2005) on chimpanzees that involved two techniques to open a box;

while they were in a completely new situation, the two females were able to teach the corresponding skill after a relatively short amount of time.

While the transmission of skills is very obvious from a semiotic standpoint, the transmission of more abstract knowledge is more complex and can often only be shown by the observation of its effects. A particularly striking example of encyclopaedic knowledge has been observed for a long time amongst chimpanzees and bonobos; what we call "the pharmacopoeia of primates". Chimpanzees and bonobos have a vast knowledge of the plants in their environment and their various curative properties that goes as far as knowing how to administer them – rubbing a leaf on the wound, swallowing the whole leaf, chewing the root or the stalk, lick the sap… Researchers (Krief et al. 2005) realised that neighbouring communities of chimpanzees and humans often had the same pharmacopoeia and used the same plants, such as the *Ficus asperifolia* with its "sandpaper" leafs, used to combat intestinal parasites.

It goes without saying that such considerable, but seldom used, knowledge (chimpanzees do not end up ill every other day) is an important investment, from both the "teacher" and the learner, that has a terrible ratio of energy expended versus utility. However, when the need to use this skill arises, it is quite often vital (malaria, fever or an infected wound can kill someone within days).

First, this teaches us about the precision of the transmitted message, as defined earlier in the chapter. The message must allow an individual to tell apart one plant from hundreds, possibly thousands of others, when the slightest mistake could be fatal. Moreover, it must explain what the plant is used for, what symptoms make it necessary and how it should be used, despite natural reluctances (some plants, such as *Verononia amygdalina*, which is useful against light infection or fever, are very bitter). Some recent studies tend to prove that chimpanzees are capable of something even more complex, combinations that allow for a more powerful and/or longer lasting effect, or that are easier to swallow or digest (*Académie Nationale de Médecine* 2011).

Secondly, this teaches us about the outlook these species have of the world, where the possibility that an individual might get sick or wounded 1 day and possibly die is important enough to their peers that they will sacrifice hundreds of hours learning these skills. This point is an important aspect of the semiotic study of animals – it is a very telling example of the possibility to reach certain phenomena that are difficult to know (in this case, the importance given to the life of a peer) through more visible signs, even if they are separated by a period of time (in this case, the important investment into education to protect oneself from disease in the future).

The fact that it is a shared support to a value or moral system from the group is another cultural characteristic. The question of whether morality exists in animals has always been a source of debate, with the reason being that it is one of the main characteristics of the *Homo* genus. Along with permanent bipedalism and brain size evolution, our species is characterised by cultural activities, that is to say social life, tools, funeral rites, art, the mastery of fire, and, of course, morals (Chapouthier 2009).

It is not possible to discuss all the possible morals aspects in a culture here, but after having studied empathy earlier, we would like to move on to an example that comes straight from it, and one of the most interesting semiotic cases, the concept of altruism.

Altruism can be considered the application, in a social structure, of the concept of empathy. It is the capability to put oneself in someone else's place in such a strong way that this feeling puts the other's well-being first, sometimes at the detriment of one's own.

Stanley Wechkin et al. (1964) set up an experiment in which rhesus macaques were trained to pull a lever to receive food. Once they understood this mechanism, another macaque was put into an adjacent cage, in their field of vision. Every time a rhesus macaque would pull the level, they would get their food, but now, the one in the other cage would receive an electric shock. Their reaction towards the lever was to quickly change their behaviour and choose to starve themselves rather than harm the other macaque; two of them went as far as not eating for 5 and 12 days.

A very interesting point semiotically speaking is that whenever this type of situation (another of their species being in danger) with chimpanzees has been observed, not only do they not want to profit from the situation, but they will also interact with the one in danger to try and help them, even at their own expense (Langergraber et al. 2017). In this case, there was two manifestations of altruism. The first was behavioural, choosing the security of another above one's own well-being, and the second was interactional, trying to help another macaque they assumed to be in danger.

Hal Markowitz (1982) devised another experiment. He trained Diana monkeys to put tokens in a machine to receive food. The tokens were distributed to each individual regularly; there was no risk of not eating one's fill but also not enough to satisfy the natural self-indulgence of Diana monkeys. Nonetheless, Markowitz saw on three occasions a young male help an old female in distress. He would pick up the token that she would drop, put it in the machine himself and give her the food, resisting three opportunities to steal the food for himself.

We can also point out the examples of Azalea (De Waal et al. 1996), a female rhesus macaque with Down's syndrome that would be fed and tolerated by the others despite being incapable of social interaction with them, of Wolf (De Waal 1997), a female blind stump-tailed monkey that would be guided and protected by the adult males in the colony, and also of Wania (Fedigan and Fedigan 1977), a Japanese macaque with a heavy neurological handicap, that would get nursed to sleep and comforted despite his screams being a risk to attract predators and his constant aggression towards the other members of the group.

Monkeys are not the only ones able to show a high level of altruism: in 1977, Porter (1977) observed false killer whales beach themselves next to their sick leader and only give the signal to leave once the leader had died.

If we wish to dig into the notion of value system a little further, at the origin of any morality is the question of what is right and what is wrong. We see in this the capital importance of empathy and altruism. These two notions can be found in the famous phrase, "do unto others as you would have them do unto you". That is the

Christian "golden rule" version but there are many instances and variations in all religions and philosophes of the world. If we want to look for the universals of morals, this might just be one of them.

To Koko, Washoe or Panbanisha, to be "good" (a "good gorilla", as Koko says) is to be nice and attentive to others. Koko shows this through her kindness to humans and her carefulness with kittens. Furthermore, she can describe an individual as "good" simply from the degree of affection she has towards them, such as when she chose Ndume as her new partner, on the videos shows by Patterson (which she expressed by signing "good gorilla really Koko love" before kissing the image of Ndume on the screen) (Brennan et al. 1999). Panbanisha defines herself as "good" on the lexigram keyboard she uses to express herself, and does not like to be told the opposite. When facing the keepers' dogs, animals she has never liked, she can be aggressive. At one point, she was abusive towards a German shepherd for no reason, and Savage-Rumbaugh told her she was "bad". After ostensibly looking away for a moment, Panbanisha pressed "good" on the keyboard twice. The researcher ended up telling her to go apologise to the dog, which Panbanisha did with gently petting it. This means that she does not only have a precise notion of good or bad, she also know in which category she wants to be, even if it means doing things she does not like, like petting a German shepherd.

Similarly, the notion of was is "bad", "wrong" or "mean" is just as well understood by the subjects. To Koko, the news of the death of All Ball is a "bad" thing because it makes her "sad", which shows she associates the concept of "bad" with an array of particularly negative notions and emotions. Washoe would sometimes happen to bite Fouts and Mills (1998) and then apologise and carefully touch his finger, and would grow more careful if it was very painful. The other chimpanzees that were with Washoe would describe breaking or stealing objects as something "bad", which shows moral judgement. Alex would call himself "bad" when he would bite Pepperberg (1999). In both cases, a particularly strong semiosis is deployed around different attitudes, behaviours and exchanges, sorted by individuals in a specific category. While these elements are fairly obvious for animals that we taught intermediary sign systems to and how to use human-animal interfaces, this semiosis is strong enough to be noticed even outside this frame.

One last element that can be found as a culture characteristic is the ritual. A ritual is a complex semiotic phenomenon that carries strong symbolism and generally a heavy emotional charge. It is a way to do something impossible to achieve (invoking rain, for example) by doing something we can do (in this case, dancing). There are many types of rituals (a rite of passage, seasonal rituals, etc.), but we only really studied one in animal studies: rituals surrounding death.

The funeral rite is without a doubt the most emblematic form of culture, as well as one of the most advanced, and indisputably the one that teaches us the most about the way those who practise it see the world. Carrying out a funeral rite on a body entails, first, being aware of the individual's death and understanding their state as different from sleep, and second, regarding this body as not just any dead flesh, but rather what is left of someone that was known, sometimes loved, and should be treated with a certain level of deference and respect.

The earlier example of the false killer whales beaching themselves next to their leader to stay with them until they died has sometimes been considered by ethologists as a funeral rite that exists in some groups of cetaceans and a potential explanation for mass beachings that can happen in some coastal areas, where animals find themselves trapped by the movement of the tides but only one of them appears to already be sick or wounded (another proposed explanation is that the wounded are disoriented and accidentally beach themselves, and the others end up trapped next to them in an attempt to rescue them). This would be similar to a funeral wake. More generally, sorrow in cetaceans takes different shapes. In 1990, researchers that had been studying the same group of orcas for 30 years noticed that two young orcas, recent orphans, were swimming apart from the rest of the group without being called back into it: they were swimming to places that they had previously gone to with their mother (Rose 2000).

Another example on funeral rites is given by Moss (1989), this time with elephants after the death of Tina, a female member of the herd. The main goal here was to understand that while the form the ritual took was the same for all involved, its meaning was not the same for all members of the group. Everyone watched over and covered the body with branches, but it was Tia and Trista, who were closer to Tina, who had the idea to gather branches and were more willing than the rest to go through this additional effort, and it was Teresia, Tina's mother, who stayed the longest during the wake and did not want to leave her daughter's body's side. This is where we reach the true meaning of the ritual: it does not have the same meaning for everyone involved, and that is how it is both a group and an individual cultural practice. The funeral rite is a visible manifestation of an invisible mourning process.

This conclusion is particularly agreed upon by primatologists. Geza Teleki (1973) describes the accidental death of Rix, a Gombe chimpanzee, killed by falling rocks. The other chimpanzees shouted, observed the body, let out plaintive cries while carefully touching him and watched over the body. His then-partner sat an entire hour next to his body without moving. Mourning in primates can be so violent it can become pathological. Tetsuro Matsuzawa's team (1997), while in Guinea, filmed Jorko, a sick two-and-a-half-year-old chimpanzee, that his sister tried to help by playing with him and hugging him, and also focused on the behaviour of Jire, his mother, who carried his body for 27 days.

While it is difficult to define culture, it is a concept that should not be pushed aside when studying animal semiosis. Not so that we can determine what is or is not cultural amongst them, but because certain manifestations, that could be qualified as cultural, are semiotically interesting in the sense that they let us know a lot about how the individuals and groups involved function.

References

Académie Nationale de Médecine – France. (2011). *Bulletin de l'Académie Nationale de Médecine, 195*(8), 1927–1944.

Armony, J., & Vuilleumier, P. (Eds.). (2013). *The Cambridge handbook of human affective neuroscience*. New York: Cambridge University Press.

Bekoff, M. (2007). *The emotional lives of animals*. Novato: New World Library.

Bekoff, M. (2009). *Les émotions des animaux*. Paris: Éditions Payot & Rivages.

Benveniste, E. (1966). *Problèmes de linguistique générale*. Paris: Gallimard.

Ben-ze'ev, A. (2010). The thing called emotion. In P. Goldie (Ed.), *The Oxford handbook of philosophy of emotion* (p. 61). New York: Oxford University Press.

Bickle, J. (Ed.). (2009). *The Oxford handbook of philosophy and neuroscience*. Oxford: Oxford university press.

Brennan, B., Malakoff, D., & Visty, R. (1999). *A conversation with Koko*. Nature PBS, août 1999.

Brown, K. (Ed.). (2006). *Encyclopedia of language and linguistics* (Vol. 1). San Diego: Elsevier Distributor.

Chapouthier, G. (2009). *Kant et le chimpanzé, essai sur l'être humain, la morale et l'art*. Paris: Belin.

Chapouthier, G., Kreutzer, M., & Menini, C. (Eds.). (1980). *Psychophysiologie. Le système nerveux et le comportement*. Ligugé: Études vivantes.

Chevalier-Skolnikoff, S. (Ed.). (1977). *Primate bio-social development: Biological, social and ecological determinants*. New York: Garland.

Christen, Y. (2011). *L'Animal est-il une personne ?* Barcelone: Flammarion.

de Saussure, F. (2002). *Ecrits de linguistique générale*. Paris: Gallimard.

De Waal, F. (1997). *Good natured: Origins of right and wrong in humans and other animals*. Harvard: Harvard University Press.

De Waal, F., & Robert, M. (2002). *De la réconciliation chez les primates*. Paris: Flammarion.

De Waal, F., Uno, H., et al. (1996). Behavioral retardation in a macaque with autosomal trisomy and aging mother. *American Journal of Mental Retardation, 100*(4), 378–390.

Dorigo, M., & Stützle, T. (2004). *Ant colony optimization*. Cambridge: MIT Press/Bradford Books.

Eco, U. (1979). *A theory of semiotics*. Bloomington: Indiana university press.

Eco, U. (1999). *Kant et l'ornithorynque*. Paris: Grasset.

Fedigan, L., & Fedigan, L. M. (1977). The social development of a handicaped infant in a free-living troop of Japanese monkeys. In S. Chevalier-Skolnikoff (Ed.), *Primate bio-social development : Biological, social an ecological determinants* (pp. 205–222). New York: Garland.

Fouts, R., & Mills, S. T. (1998). *L'école des chimpanzés: ce que les chimpanzés nous apprennent sur l'humanité*. Paris: J. C. Lattès.

Goldie, P. (Ed.). (2010). *The Oxford handbook of philosophy of emotion*. New York: Oxford University Press.

Goodall, J. (1986). *The chimpanzees of Gombe*. Cambridge: The Belknap Press of Harvard University Press.

Greimas, A. (1988). De la nostalgie : étude de sémantique lexicale. In *Hommage à Bernard Pottier vol. 1* (p. 343). Paris: Klincksieck.

Greimas, A., & Fontanille, J. (1991). *La Sémiotique des passions. Des états de choses aux états d'âme*. Paris: Seuil.

Gucwa, G., & Ehmann, J. (1985). *To whom it may concern: An investigation of the art of elephants*. New York: W. W. Norton & Co.

Jakobson, R. (2003). *Essais de linguistique générale, 1: Les fondations du langage*. Paris: Éditions de Minuit.

Konishi, M. (1965). The role of auditory feedback in the control of vocalization in the white-crowned sparrow. *Zeitschrift für Tierpsychologie, 22*(7), 770–783.

Krief, S., Hladik, C. M., & Haxaire, C. (2005). Ethnomedicinal and bioactive properties of plants ingested by wild chimpanzees in Uganda. *Journal of Ethnopharmacology, 101*, 1–15.

Langergraber, K. E., Watts, D. P., Vigilant, L., & Mitani, J. C. (2017). Group augmentation, collective action, and territorial boundary patrols by male chimpanzees. *Proceedings of the National Academy of Sciences of the United States of America, 114*(28), 7337–7342.

Maestripieri, D. (Ed.). (2003). *Primate psychology*. Cambridge: Harvard University Press.

Manser, M., Bell, M., & Fletcher, L. (2001). The information that receivers extract from alarm calls in suricates. *Proceedings of the Royal Society B: Biological Sciences, 268*(1484), 2485–2491.

Markowitz, H. (1982). *Behavioral enrichment in the zoo*. New York: Van Nostrand Reinhold.

Marler, P., & Mundinger, P. (1971). Vocal learning in birds. In H. Moltz (Ed.), *Ontogeny of vertebrate behaviour* (pp. 389–450). New York: Academic.

Matsuzawa, T. (1997). The death of an infant chimpanzee at Bossou, Guinea. *Pan Africa News, 4*(1), 4–6.

Moltz, H. (Ed.). (1971). *Ontogeny of vertebrate behaviour. New-York*. Academic.

Morris, D. (1962). *Biologie de l'art*. Paris: Stock.

Moss, C. (1989). *La Longue Marche des éléphants*. Paris: Robert Laffont.

Mounin, G. (Ed.). (2004). *Dictionnaire de la linguistique*. Paris: Presses Universitaires de France.

Naguib, M. (2006). Animal communication: Overview. In K. Brown (Ed.), *Encyclopedia of language and linguistics* (Vol. 1, p. 277). San Diego: Elsevier Distributor.

Nishida, T., et al. (2010). *Chimpanzee behaviour in the wild: An audio-visual encyclopedia*. Tokyo/New York: Springer.

Parr, L. A., & Maestripieri, D. (2003). Nonvocal communication. In D. Maestripieri (Ed.), *Primate psychology* (pp. 324–258). Cambridge: Harvard University Press.

Pepperberg, I. (1999). *Alex studies. Cognitive and communicative abilities of grey parrot*. Cambridge: Harvard University Press.

Porter, J. W. (1977). Pseudorca stranding. *Oceans, 10*, 8–15.

Prinz, J. (2010). The moral emotions. In P. Goldie (Ed.), *The Oxford handbook of philosophy of emotion* (pp. 519–538). New York: Oxford University Press.

Pugmire, D. (2010). Saying it. In P. Goldie (Ed.), *The Oxford handbook of philosophy of emotion* (p. 384). New York: Oxford University Press.

Rose, N. A. (2000). A death in the family. In M. Bekoff (Ed.), *The smile of the dolphin. Remarkable accounts of animal emotions* (p. 144). New York: Discovery Books.

Sperber, D., & Wilson, D. (1989). *La pertinence: Communication et cognition*. Paris: les Éditions de Minuit.

Stoeger, A. S., Mietchen, D., Oh, S., de Silva, S., Herbst, C. T., Kwon, S., & Fitch, T. (2012). An asian elephant imitates human speech. *Current Biology, 22*, 2144–2148.

Sufka, K., Weldon, M., & Allen, C. (2009). The case for animal emotions: Modeling neuropsychiatric disorders. In J. Bickle (Ed.), *The Oxford handbook of philosophy and neuroscience* (pp. 522–536). Oxford: Oxford University Press.

Teleki, G. (1973). Group response to the accidental death of a chimpanzee in Gombe National Park. *Folia Primatologica, 20*(2), 81–94.

Thoen, H., How, M. J., Chiou, T. H., & Marshall, J. (2014). A different form of color vision in mantis shrimp. *Science, 343*(6169), 411–413.

Watanabe, S., & Kuczaj, S. (2013). *Emotions of animals and humans – Comparative perspectives*. Tokyo: Springer.

Wechkin, S., Masserman, J. H., & Terris, W. (1964). Shock to a conspecific as an aversive stimulus. *Psychonomic Science, 1*, 47–48.

Whiten, A., Horner, V., & De Waal, F. (2005). Conformity to cultural norms of tool use in chimpanzees. *Nature, 437*(7059), 737–740.

Wittgenstein, L. (1961). Investigations philosophiques. In *Tractatus logico-philosophicus suivi de Investigations philosophiques* (p. 1961). Paris: Gallimard.

Zahn, R., de Oliveira-Souza, R., & Moll, J. (2013). Moral emotions. In J. Armony & P. Vuilleumier (Eds.), *The Cambridge handbook of human affective neuroscience* (pp. 491–508). New York: Cambridge University Press.

Zuberbühler, K. (2013). Referential signalling in non-human primates: Cognitive precursors and limitations for the evolution of language. *Advances in the Study of Behavior, 33*, 265–307.

Chapter 4
Semiotic Tools and Concepts

Abstract Conception sub-question n°3: what are the semiotics tools and concepts that are interesting and desirable to work with in animal studies?

First, this chapter presents the semiotic tools that are interesting to work with in animal studies to describe non-human semiosis and how to choose them. The presented tools come from the relevance theory (Sperber and Wilson version), Peirce's theory of signs, but also from our own work regarding the IFC (Intensity, Frequency, Context) grid. Secondly, this chapter will introduce the semiotic concepts with which to interpret semiosis described with the help of the tools.

The chapter will thus present the grid of Intentional-Conscious- Unconscious signs, the interpretative use that can be done of the phatic function of speech of Jakobson, but also the way some semiotic concepts by Eco can help with interpretation.

This chapter will end with the specific case of anthropomorphism and the question of whether it can be used as a tool.

Keywords Semiotic tools · Semiotic concepts · Umberto Eco · Roman Jakobson · Charles Peirce · Relevance theory

Now that we have cleared the field of used notions, we can focus on the heart of methodological models: the tools and concepts. In this chapter, I will finish what more or less constitutes the preliminary step to my methodology, that is to say its framework, what it is a part of (the debates, the definitions and now, the tools).

This chapter will therefore be a way to answer the last conception sub-question from the introduction: What are the semiotic tools and concepts that are interesting and desirable to work with in animal studies? Tools and/or concepts are considered interesting regarding this methodological work when they allow for a better layout for the model, when they are particularly relevant in animal studies and when they are somewhat flexible, making it possible to adapt them to various species that a semiotician will work with. These same tools and/or concepts are considered desirable when they easily allow a bridge between two disciplines, when they are not

P. Delahaye, *A Semiotic Methodology for Animal Studies*, Biosemiotics 19,
https://doi.org/10.1007/978-3-030-28813-6_4

subject to the controversies and debates from Chap. 2 and they can be functional while rarely relying on concepts which have problematic definitions as seen in Chap. 3.

In the first part of the chapter, I will focus on semiotic tools, specifically the selection method of our methodology. How should a semiotic tool for a study on animals be chosen? What should guide our choice and how can we know our choice is the best possible one? I will of course discuss the reading grid, which is particularly useful in this type of work, the tools made available for researchers by Peircean semiotics (and why I favour this school of thought in animal studies), as well as a grid of signs I established for our thesis work called IFC, for Intensity, Frequency, Context.

In the second part of the chapter, I will focus on semiotic concepts that are not really tools for capturing, sorting or weighing signs, but rather means to interpret them in a coherent unit. I will talk about the triple concept of Intentional-Conscious-Unconscious which allows me to interpret the voluntary aspect of "saying", the special case of the phatic function of language according to Jakobson (2003), the semiotic theory of Eco and the interpretation it offers in zoosemiotics, and I will finish this chapter by the special (and problematic as far as the criteria of "desirable concepts" is concerned) concept that is anthropomorphism.

4.1 How to Pick Semiotic Tools

In this part of the chapter, I will go into the details of the semiotic tools I will use to create my methodology of semiotic study of animals. However, I will only discuss the semiotic tools proper. There are other tools, some coming from ethology for example (ethograms are a noteworthy example) but I will discuss them during our next chapter, when I will discuss how to build a methodology with tools from different academic fields.

I will put a particular emphasis on the way these tools are chosen and why, as well as the way to apply them to fields and topics that sometimes seem very distant from their original discipline.

This is why I will naturally begin by speaking about the theory of relevance, which in this model will be a sort of meta-tool, a tool that lets us select, sort and position all the others. Relevance will also have a significant importance when joining with life sciences and their tools and methods (I will return to this in more details in Chaps. 5 and 6).

There have been several forms of the theory of relevance, from different authors (the works of Grice (1982) or Fodor (1983) on the topic come to mind), but I will focus here on what seemed to us to be the most complete form of this meta-tool, which comes from Sperber and Wilson.

Then I will discuss the tools from Peircean semiotics I will be using, as it is obvious that some of Peirce's tools and methods are too closely connected to textual analysis or strictly human productions to be adapted to animal studies without completely distorting the applied concepts. I will mostly explain why I choosing these tools is relevant to study animals, and will not do a detailed description of the tools in question, as others have done it much better than I could already (Deledalle 1971).

At the end of my review, I will introduce the IFC sign reading grid developed during our thesis work. I will explain why I chose these characteristics, why these criteria are important in animal studies and what this grid brings to the table regarding semiotic work on the topic. The examples given will mostly be related to the semiotic analysis of emotions, as it was the theme of our thesis when this tool was built, which is why the illustrations are cases I were working on at the time.

4.1.1 Relevance

Sperber and Wilson are not the first to have looked into the concept of relevance and have created theories on the subject, Paul Grice (1982) or Jerry Fodor (1983) can also be mentioned. Their works are different from Sperber and Wilson's and, while they are not a good fit for animal study so to speak, we are discussing them for the benefit of semioticians that wish to work on the way the mind works.

The dictionary definition of relevance is "the quality of what is logical, perfectly appropriate[1]". However, the way Sperber and Wilson (1989, p. 189) describe it shows broader implications:

> On pourrait comparer le concept de pertinence avec des concepts comme ceux de productivité ou de rendements, concepts qui s'analysent en termes de coûts et de bénéfices. [...] Les processus mentaux, comme tous les processus biologiques, demandent un certain effort, une certaine dépense d'énergie. [...] L'effort de traitement est un facteur négatif : toutes choses étant égales par ailleurs, plus l'effort de traitement est grand, plus la pertinence est faible.
>
> Translation:
>
> We could compare the concept of relevance with concepts such as productivity or output, concepts that are analysed in terms of costs and benefits [...] The mental processes, as any other biological process, demand a certain amount of effort, a certain expense of energy. [...] The treatment effort is a negative factor: all things being equal otherwise, the bigger the treatment effort, the weaker the relevance.

Relevance theory aims not to describe what happens when an individual expresses themselves, learns or thinks, but rather to explain how these mechanisms function and why a specific cognitive or linguistic "strategy" was chosen above another.

This theory can be used in particular to determine the potential reasons for failure (or unexpected results) of certain tests. I have already mentioned the mirror test earlier, and the reasons that could lead one to understand why certain species like

[1] *Le petit Larousse illustré 2010,* Paris, Larousse, 2009.

dogs and cats fail it (or appear to, as we previously discussed) while magpies, for example, are successful.

First of all, this theory is necessary to gauge the degree of ecology in experiment or study reports I wish to analyse, and detect a potential bias that could call the conclusions from these experiments into question.

Secondly, relevance theories will help us take into account the different parameters that can come into play in the theoretical model. Are we talking about an individualistic or social species? What is its dominant sense, its privileged information channel? What is its life expectancy, and can it learn throughout all of it? These questions that are seemingly unrelated to our issue are actually just what will form the model's ecology. It is most likely more relevant for a social species to develop tightly knit relationships and an elaborate mode of communication than it is for a species whose members live relatively isolated from one another. Similarly, the dominant sense or information channel can explain the origin of this or that mode of communication (for example, what would be the point for an aquatic species to develop an articulated language?). Finally, the length of a species' life expectancy can result in the creation of emotional links of different strengths.

As we can see, relevance is mostly a "sorting" tool. While its usefulness does not stop at this one function, it is the one we will explore first and foremost. It will first allow us to select which tools, methods and concepts use in which situation.

Through these questions, this permanent analysis of what is or is not a cost too high for the subject in terms of mental process, relevance can be used as a protection against anthropomorphism, particularly by stopping us from projecting onto situations some elements that are not there and, according to the knowledge we currently have, cannot be there.

In a second step that almost mirrors the first, relevance can be used to remove any trace of anthropomorphism from situations presented to a researcher (observation, experimentation, study reports, etc.) by flushing out irrelevant causal links, hasty conclusions or other possible explanations of a phenomenon that should have been dismissed. The goal here is to avoid the second trap I mentioned in Chap. 1, that I see as the most harmful due to its consequences but also as the one we have the fewest methodological tools to avoid.

Lastly, it allows us to give precise criteria to identify the cases where we end up with a study that has inappropriate (conceptual or experimental) tools, as well as the way to remedy this to get convincing results, valid protocols or relevant concepts.

The use of relevance should not be seen as an attempt to distort existing concepts but rather as an attempt to take into account the subjective part they are built on and how to deal with it (Hoffmeyer 2008 p. 175):

> [...] the scientific community has been little influenced by the obvious consequences of this insight, namely that scientific reality itself is a human, and therefore humanly limited, construction. Or, to put it in the famous quote from Einstein, "Scientific concepts are free creations of the human mind, and are not, however it may seem, uniquely determined by the external world" (Einstein and Infeld, 1938).

This idea of concepts dependent upon the subjectivity of the human mind is coherent with the taking into account of experimenter biases discussed in the first chapter.

How can the correct use of relevance help us avoid these two major traps, and ensure we have the right tools in the framework of a study on behaviour, intelligence or animal emotion?

First, it is necessary to be well-acquainted with the biology of the studied species; how it lives, how it sees, how members of the species interact. In a practical study, it should always be checked whether the proposed tools are conform to the individual's motor skills, sense and social habits.

Let us discuss some examples. It is obvious that the motor skills of the *corvidae* are not the same as the primates'. While the latter have hands very similar to ours, the creation of tools for testing a magpie's ability to carry objects, for example, needs to take into account the morphology of its beak and the maximum weight it can handle as well as the fact that some puzzles might be unsolvable due to the bird' inability to reach certain parts as its head can only rotate up to a point, which would not be a problem with hands like ours.

Another example I have previously mentioned is the case of the mirror test used on a cat, which creates the same type of problem. I have already shown how this test is hardly relevant considering the biology of cats, but did not offer a suitable solution. At this time, we have yet to find a "mirror test" protocol applied to the sense of smell. We are still struggling to find test that call to other primary senses, resulting in many species deprived of tests relevant to their biology.

Finally, the example we gave in Chap. 1 about the meta-analysis of Boesch (2007) is very telling regarding the impact that social ties can have: while it varies from species to species, forgetting that the individual is sometimes first and foremost a member of a group with strong social interactions takes away a part of its usual way of functioning and greatly twist the results of the experiments.

As for the case of unknown biological elements – as is often the case with emotions, which can be a discrete and internal phenomenon – relevance allows us to apply the elements from one phylogenetic branch to another in order to try and find similitudes. Between phylogenetic branches. An ability that is very present in one or several species of the same branch is obviously relevant to their functioning and can be suspected to exist in related species. In this situation, I take the abduction method first explained by Peirce then Eco (1988) which lets us study signs that "we suspect to be signs, without knowing what they are signs of" (Eco 1988, p.51).

Let us take an example from our close cousins again. The existence of sorrow and grief in the majority of humans no longer needs to be proven (and may never have needed to be proven), each of us unfortunately experiencing these emotions at some points in our lives and expressing them in very different ways.[2] Similarly, Fouts and Mills (1998) brought to light the behaviour of sorrow caused by death,

[2]To take a literary example, Camus' character, Meursault, is criticised for not crying at his mother's funeral, showing that even amongst humans, sorrow and grief can be expressed in different ways.

grief and even pathological grief during his years of studying Washoe, whose species belongs to the *Pongidae* family, alongside ours (Grassé 1955). Based on these observations, it is easy to understand why Patterson (1986) and his teams interpreted Koko's reaction to learning about the death of All Ball as the expression of grief, as gorillas also belong to this family, despite the fact that the existence of this phenomenon had yet to be proven in that species.

Lastly, relevance enables us to consider the individual and to remind ourselves that an animal of a certain species is a complex entity with its own history, distinctive features and character, and is not exactly the same as another animal from the same species (we owe a lot to primatologists, Goodall primarily, for having established this notion in ethology). Keeping this notion in mind leads to avoiding mistakes or the misunderstanding of results due to individual parameters that would have not been taken into account while setting up the experiment. Anyone living with two cats, dogs or rats simultaneously will notice how the characters and intelligences of the two individuals differ, sometimes evolving in opposite directions.

The elephants of Nissani (2006) and their test on causality is a striking example. While it is doubtlessly preferable to administer this type of test to domesticated elephants used to the presence of people, making an animal accomplish certain tasks when they are trained (often through violence) to do strictly and exactly as they are asked at the risk of being immediately punished is relatively unconvincing/questionable.

Subject choice is a primordial aspect of behavioural experiments and neglecting the possible impact of the environment and past experiences of an individual exposes researchers to adding considerable biases without even realising it. The researcher must ask themselves if the experiment is sensible to do with the specific animal in front of them, if it is relevant for their mental world or if, on the contrary, the experiment is completely subjected to another phenomenon that conceals the subject's conscience while not preventing the completion of the experiment, leading to confusion within the results.

The relevance theory is also a tool that will enable us to adapt our semiotic concepts to the different cases we will encounter during a study on animals. I have centred this part of the building of my model on three major "dual situations" that can be encountered in ethology and strongly impact a model.

Firstly, we must look at the "social – solitary" pair, certainly a crucial pair of variations in a study like mine on emotions that mainly relies on their observable expression. Within this scope, the relevance theory will help us understand at which moment a behaviour becomes indicative of signification (we could almost say signifying) and, on the opposite side, in which cases its non-observation will be (or not be) conveying meaning.

Secondly, we must direct our attention towards a pair that is already regularly mentioned by ethologists, the "freedom – captivity" pair. It is well-known by observers of the animal kingdom, and I simply want here to remind how the same event can be relevant and meaning carrier in an animal in captivity and be neither of those in animals of the same species living in nature, and vice-versa. But this study is also a reminder of what remains relevant in all circumstances in members of the same species, and how we can and must be careful not to dismiss a manifestation too quickly on the grounds that an individual is, for example, in captivity.

Lastly, we should pay attention to a coupling that still often goes unnoticed, the "group – individual" pair. There are three reasons to apply relevance theory to this coupling. The first reason is reminding that the results obtained from an isolated individual cannot always be applied to all members of their species. The second is understanding that despite this, a result obtained from an individual must make us suspect the existence of a similar process in other members of their species, even when it comes to rare events. The third reason is being aware of the fact that as soon as we deal with species that are capable of feeling, intelligent or conscious, the group shows a general tendency, a standard behaviour but large disparities between specific individuals can exist and can only be explained by the individuals' own, distinct identities.

This notion of adaptability is essential to me. As Kalevi Kull (2012, p. 18) explains, we must take into account the possibility of creation, unpredictability and a certain ambiguity, which are all specific to the semiosis.

> Semiosis is what is capable of creating new code-relations. Simultaneously, semiosis also carries on existing codes, rebuilding and inheriting these. Semiosis always includes certain code. Code is a necessary but not a sufficient condition for semiosis. [...] The capacity of creating a new code implies that semiosis is also a unit of learning from experience. This means that semiosis assumes certain ambiguity, certain indeterminacy, unpredictability.

This notion of adaptability linked to relevance theory will be completed, amongst other things, by tools linked to Peircean pragmatism that also are tokens of adaptability.

4.1.2 Peirce's Tools

Peirce (1987) is without a doubt one of the theorists that has contributed the most to the study of the complexity of sign components. By combining a scale of form complexity of the sign (qualisign, sinsign, legisign) with another scale of complexity of logical link to the represented object (icon, index, symbol) and another of complexity of the signification process (rheme, dicisign, argument), he gave semiotics a priceless tool of analysis for signs and their intrinsic complexity (Peirce 1992). This complexity has sometimes been misunderstood or qualified as being too metaphysical. However, as Favareau (2006, p. 30) very well puts it:

> The point is that there is nothing mysteriously "metaphysical" about Peirce's notions of what he calls **firstness**, **secondness** and **thirdness**. Rather, these relations refer, in a radically necessary relation of **possibility**, **existence** and **law**.

I had already studied at length during my master's dissertation how Peircean classification can help understand mental mechanisms in species other than ours (Delahaye 2013). I include below (Table 4.1) as a reminder a summary of Peircean classification.

Table 4.1 Semiotic classification of Peirce

Category	Complexity		Short definition	Compatibility
Representamen	1	Qualisign	A quality that functions like a sign	Icon – Rheme
	2	Sinsign	A specific spatio-temporal thing or event that functions like a sign	Icon/Index – Rheme/ Dicisign
	3	Legisign	A conventional sign	Icon/Index/Symbol – Rheme/Dicisign/ Argument
Object	1	Icon	A sign that resembles the object	Qualisign/Sinsign/ Legisign – Rheme
	2	Index	A sign affected by its object and that reveals it (fever, footsteps, etc.)	Sinsign/Legisign – Rheme/Dicisign
	3	Symbol	A sign referring to its object through code and often arbitrarily	Legisign – Rheme/ Dicisign/Argument
Interpretant	1	Rheme	A sign on its own, without context, an isolated name, interpreted through its simple existence	Qualisign/Sinsign/ Legisign – Icon/Index/ Symbol
	2	Dicisign	A sign in the context of a logical proposition that may be true or false	Sinsign/Legisign – Index/Symbol
	3	Argument	A sign in the context of a reasoning through deduction, induction or abduction	Legisign – Symbol

Table 4.2 Illustration of the classification of signs of Peirce

Semiotic combination	Example
Qualisign – Icon – rheme	A colour, a smell, a vague feeling
Sinsign – Icon – rheme	A realistic drawing, a non-symbolic pictogram
Sinsign – Index – rheme	A cry of pain, a jolt, a stutter
Sinsign – Index – dicisign	Pointing at an object, following a track on the ground
Legisign – Icon – rheme	An onomatopoeia, an animal mimicking a human voice
Legisign – Index – rheme	A territorial scream, a generic warning shout
Legisign – Index – dicisign	A word in sign language that mimics the real object: Imitating the fluttering of wings with one's hands for « butterfly »
Legisign – Symbol – rheme	An isolated name, a specialised warning shout
Legisign – Symbol – Dicisign	A grammaticised sequence of words: « Penny give Koko juice » is different from « Koko give Penny juice »
Legisign – Symbol – Argument	A logical proposition

For more clarity, I also provide below (Table 4.2) some examples of classification of signs observed in my work:

In his classification, Peirce included amongst the simpler signs the qualisigns such as inarticulate screams. They have often been associated with the cry of pain, for example, an involuntary scream if there has ever been one, which clearly signified the physical pain felt by an individual without saying much about their train of thought at that exact moment.

But what about the cry of "mental" pain? Individuals – including humans – frequently let out an inarticulate scream when dealing with enormous anger or sorrow. It is not as involuntary as the cry of pain, and the individual will often try to either restrain it or freely let it out depending on the external circumstances. It is a precious indicator of the state of emotional "saturation" of the mental world, that is to say the emotion at the source of the scream so largely occupies the subject's mental world that, one on hand, it is difficult for them to contain it and often requires to be expressed so as to be relieved – more or less, depending on the case – but also, on the other hand, it completely parasitises their cognitive processes and does not give them the choice of a more complex and elaborate expression. Therefore, a simple sign is not so much the mark of a non-complex emotion as much as it is the mark of the intensity of said emotion.

Other signs can also belong in relatively simple categories in Peircean classification while still being indicators of complex emotion and thought mechanisms.

For example, the case of the name as it is used in dolphins to refer to signal oneself to others in their group and call each other is a relatively simple sign that can be used outside of any context as it only has a "labelling" value (it is a rheme, the first notch of the last Peircean scale). But is it an indicator of a mechanism that predisposes individuals to complex emotions: the individuality of others. Considering another member of the group as a unique being naturally predisposes individuals to feel emotional attachment, and through a knock-on effect, violent emotions if that connection is broken (Béata 2013). The presence of signs of this category must make us look for others, including two that I will detail here.

The first one is the use of symbols (a complex sign) when maintaining that connection between individuals. Their presence shows the importance those connections have for the individuals involved, to the point that their group possesses a symbolic system solely to maintain them.

In this category, I can cite the cases well documented and studied in depth by De Waal and Robert (2002) of mechanisms of reconciliation in great apes, where objects are often used as carriers of an emotional message. The bringing of an offering (food, toy, etc.), its degree of symbolism (favourite food, a toy that was the reason for a dispute, etc.) are all symbols of an attempt at reconciliation, and the acceptance or reject of that offering are not so much linked to the object itself as much as to the message the individual is attempting to convey (returning the affection, anger still being present, etc.). This very complex process shows the amount of importance those species have for their connection with the other members of their group, for the quality and length of that connection. It is also possible to consider it a proof of the existence of affections so strong they can resist multiple disputes, quarrels or spats of anger.

The second sign will just be mentioned here. It is the complex sets of signs, indicators of complex and contradictory emotions in the individual who emits them (jealousy is a very good example). I see this case as under the umbrella of sign combinatorics and will be discussed below.

As explained earlier, Peircean semiotics establishes a very tight-knit link between thought and sign, as "there is no thought without sign" (Peirce 1987).

Therefore, it can be surprising to notice that the production of complex signs from individuals of various animal species has often been categorised alongside instinctive manifestations, even sometimes the completely unconscious manifestations, or at least manifestations that were not thought through, not thought at all, and not voluntary. While the theories of Peirce did go from thought to sign ("there is no thought without sign"), it is surprising the opposite direction (from sign to thought) was never really considered.

Evidently, not every sign the result of a thought. Each living being overflows with signs they produce despite themselves, sometimes simply by being alive (heartbeats, eye movements, respiration noises, etc.) sometimes out of instinctive reflex (cries of pain, turning their head due to a sudden noise, etc.). But we should wonder whether it is possible for complex signs to exist without being the result of a thought, or a cognitive or emotional mechanism that is also complex.

If we stay within was Peirce's works teach us on the sign complexity scales, it seems highly improbable that certain sign categories can be emitted by an individual without thought preceding the emission. The legisign, symbol, rheme, dicisign and argument seem difficult to imagine as being expressible without thought, without referring the sign to the object, without connecting the signs with each other (we have mixed feelings regarding the sinsign, therefore we chose not to include it for now).

Following this idea, when a researcher is confronted with an individual emitting signs, they should ask themselves this first question: which category of emitted signs are we witnessing? If they are complex signs, semiotic logic says it is not relevant to consider then as anything but coming from a complex internal mechanism. In this way, semiotic logic distances itself from Morgan's cannon, and this is the kind of legacy or tradition in a field that can be beneficial to interdisciplinary work; we will return to this point in Chap. 6.

We are then confronted with another question. What is the complexity of an individual, a group, a species? What does the sign that was intercepted by a researcher tell us about this complexity?

These questions are not the result of an "over-analysis" but rather the understanding of a real difficulty in semiotic work, especially for those of us working on emotions, which vary so wildly from one individual to the next. While semiotics is seen as "the life of signs in social life" since de Saussure (2002), it also is the life of signs in an individual's mental life. We must be careful with our interpretation, as it tries to give signs a meaning that was given by a mind other than ours when it was emitted (this problematic is identical, if less complex, in human linguistics).

Those of us trying to understand often complex internal processes through the reading and interpretation of signs must always be aware of this point. This is where the superposition of various reading grids takes on its full meaning: Peircean pragmatism enables the study of emitted signs, relevance gives us the meaning they can have in an individual's mental world, biology tells us about their sensory organs, cognitive abilities and natural restrictions that govern their life, and Peircean semiotics, again, allows us to put into a new perspective the meaning of those signs in the biological life of the individual, the group and the species.

Peircean tools can also be particularly useful on one last aspect. The notion of motivation of sign production can appear ambiguous in the sense that it is tough to know the extent in which motivation is an origin or a will. In this rest of this chapter, we will therefore use the term "motivation" in its most generic term, that is to say the trigger for an event, and the term "will" for any conscious, desired or thought-through motivation to a signifying action.

Some theorists (Swan 2012) also mention intentionality, arguing that its existence can be deduced through logic, and oppose other writers that believe it is incorrect to think there is such a thing as animal will:

> But there's an even more obvious point to consider, which is this: Why would humans be the sole possessors of intentionality in the natural world if we are the latest result of an extremely long process of active, experiencing and adapting living things? In this light, it makes no sense at all to have to **pretend** that other creatures have intentionality, no matter your philosophical position. (Swan 2012 p. 232)

I think we can list three types of motivations of sign production: voluntary motivation, involuntary motivation, and unconscious motivation.

Voluntary motivation is most likely the most obvious one to see and interpret. A semiotic act can be considered as voluntarily motivated when the subject emits signals ostensibly, going as far as calling out and looking for another individual of their group so they can witness their emotion. Actions dictated by jealousy, blatant aggressions, sulking, rarely occur, by definition, outside of the presence of the individual responsible for the issue.

It is also possible to add to this category the extremely codified emotional manifestations such as grieving rituals (Moss 1989) or reconciliation rituals (De Waal and Robert 2002). Indeed, it is highly unlikely that actions so precise, complex, coherent with each other and systematic can exist in a species or a group without being thought individually by the subjects involved. The question of the meaning of each "sub-action" of the semiotic action of an individual in a codified ritual is probably too complex to handle here; we will simply say it is likely that each subject finds their own meaning in it, as is the case in our species.

Involuntary motivation requires a little more subtlety to grasp, but specific marks do exist. They are involuntary but conscious semiotic acts, and the subject may try to hide or offset them with other signs when they realise what they are doing (Bohler 2007). These forms of motivation are what lead the most often to what we will describe as chaotic sign sets as the subject emits voluntary and involuntary signs that contradict each other.

Some forms of expression of jealousy or anger can be put in this category: strong emotions that tend to take over the individual and make them emit plenty of involuntary sign, while also having a majorly negative and stressful impact on the group.

Therefore, when they notice that negative effect, the subject will try to offset it and voluntarily emit contradictory signs.

More generally, it is unlikely that an involuntary but conscious manifestation remains isolated. In most cases, voluntary manifestations in agreement or contradiction with the first signals can be expected once the subject realises what they have been expressing involuntarily.

Unconscious motivation is without a doubt the hardest one to identify. It is particularly difficult to determine whether an emotional manifestation is conscious and accepted, maybe even voluntary, and whether it is unconscious.

If we return to the example of the grieving process I mentioned earlier, it is easy to see how it can be problematic. While a specific and precise ritual like the ones cetaceans or elephants can have is most likely a voluntary emotional manifestation, the same cannot be said of all the elements it consists of. The wake, or just as impressive, the "minute of silence" observed in ravens (Fleming 2009), for example, can be considered a voluntary manifestation while also being a completely unconscious manifestation, closer to a state of astonishment due to great mental distress.

4.1.3 Intensity, Frequency, Context

My first "adjusting tool" of the grid is without a doubt one of the most complicated to clearly perceive. A telling example is the difficulty encountered by medical staff to establish reliable and universal pain scales. While we lack the ability to measure the intensity of a process (emotion, cognition, etc.), we can measure the intensity of its expression. It does not give us direct information on the process itself, and the measurement must systematically take into serious consideration the personality of the individual (a fit of anger from a habitually aggressive individual does not have the same impact or meaning as one from a usually calm individual who would usually stop at a warning at most).

First, let us look into the "pure" intensity of a semiotic manifestation. How is it measurable?

A large number of works have for example listed in humans or other specific species – primates, dogs, cats, (Massal and Lachapèle 2014) etc. – the distinctive signs of anger, fear and sadness. It should be noted that many signs are additional, that is to say they are not all present at the same time but overlap each other depending on the emotional state of the subject: the more intense the state, the more signs from the manifestation. When a researcher wishes to study a specific species, in the case of emotion, it is recommended they take the corresponding "reading grid" of emotional signs and establish a scale of intensity in manifestations based on the number of signs being shown, as they reveal the place the emotion takes in the mental world of the subject (Béata 2005).

The intensity of a process can also be measured by its ability or lack thereof to cause exceptional behaviours. Fouts and Mills (1998) tells the story of Washoe, who had always been relatively scared of water – a completely justified fear as chimpanzees are notoriously bad swimmers – and yet one day hurriedly climbed over the fence of her enclosure and jumped into the nearby river without hesitation to save another young female that was drowning. Fouts says the cries of distress from the female led Washoe to a state of extreme agitation that she had still not left when back on the riverbank with the other female (who was safe and sound but would not survive her second drowning in the same river, and her body would never be retrieved by Fouts).It is always tempting to talk about instinct rather than an emotional or cognitive reaction when an individual acts in such a spectacular and visibly little thought way. It however seems rather unlikely that instinct alone would lead an individual to do something that so obviously goes against their most elementary survival instinct, and I think it is more probable that the situation creates such an intense and unbearable emotion in the individual (here, the dread of seeing a member of their own species die, maybe even the projection of their own sorrow if that did happen, or how distressed they would be in that happened to them) that they will do anything they can, completely disregarding the risks, to escape this emotional state – in this case by saving the struggling individual they are worried about.

Lastly, there is the less specific case of varying intensity depending on the individual. It is well known amongst humans; it was very well documented in primates by Goodall since the beginning of her research, and it has started to appear in recent years in veterinary sciences. Separate individuals, within a species or even the same group, are not sensitive to the same emotions. Goodall would mention the "soft", "prone to anger" or "playful" temperament of certain chimpanzees, and it is certain that this influences the "real" intensity of their emotions. As previously mentioned, a fit of anger from a badly-tempered individual is not the same as one from a calm one, and it is therefore sensible to suppose the latter must have been pushed very far to lead to this manifestation. Intensity can be seen as a variable that also possessive a scale that is proper to each individual and that must be understood so that the interpretation made by the researcher is as accurate as possible.

The measuring of frequency of behaviours, on the other hand, has long been an important element to ethologists. It allows them to understand the standard functioning of a species or a group and to notice the important activities that sets the tempo of their lives. I will look at this from a slightly different angle.

First, frequency can be the sign of voluntary behaviour, over an accidental or opportunistic one. That is the conclusion that Pruetz (2011), the primatologist, reached when studying the frequency at which a young male chimpanzee would go assist a wounded female and her young. A one-time occurrence could have been a non-calculated act (the young male is inexperienced and had no idea of the amount of fatigue that comes with carrying a baby for hours) or opportunistic (in the hope of receiving sexual favours or a bigger portion of food from the female), and would not have happened again once the subject had fully understood the difficulty that this assistance entails or the inability of the wounded female to reward him at all. But Pruetz noted the consistency with which the young male would help the female and the frequency of him holding the baby, the length of that action, the number of instances of assistance per day and the overall period during which this lasted.

The analysis that Claude Béata (2013 pp. 179–181) makes of this, is that this is very much a case where the frequency of assistance is such that it can only be a voluntary act fully assumed by the male despite all the constraints it entails. The same act, isolated, would most likely not have had the same meaning, and that is why we see frequency as an "adjusting tool" that allows a better understanding of what a specific behaviour really entails in the mental world of a subject: the same sign with a different frequency does not have the same meaning for the subject themselves or the researcher observing them.

In social species, frequency is a tool that can also help with evaluating the strength of the various social ties of an individual and classify them. While it is not surprising to see particular, more attentive and more frequent social ties between mothers and their young, it is interesting to study the preferences (or aversion – a subject can be aggressive by nature but the frequency of aggressions can show a "preferred target") that some adults show towards others. This is particularly true in species that permanently live in groups. Those that occasionally have groups but have permanent couples (as is the case with *Corvidae*) will of course show preference to their partner while in a group, which does not teach us much.

Similarly, in pets, it is not rare to see that the house cat or dog will show a marked preference for one of the members of the family, who will be solicited the most for playing or petting, and will be the one that the pet tends to sleep next to at night or goes to when the pet is sick or anxious. Of course, this does not mean the pet is actively avoiding the rest of the family (except in rare cases of particular dislike) or does not ask attention of gestures of affection from them, but the frequency reveals the privileged member of the family – usually the one they have known the longest, or that cared for them when they were young – with whom the relationship is different.

Lastly, frequency also enables noticing a large change in personality, behaviour or lifestyle in an individual. A behaviour or isolated manifestation do not have the same meaning as they would if the phenomenon suddenly became a regular activity for the subject and an integral part of their functioning (we could call it their new functioning). The first example that comes to my mind is the pathological example. While the change of attitude due to a grave or traumatising event is completely normal, its duration is what will determine whether a pathology remains. Koko and Washoe both reacted by being distressed then prostrate as a result of the death of an individual they were attached to. However, Koko returns to her normal life after a few days, while the behaviour of Washoe will worry Fouts and Mills (1998) for weeks.

I can also mention the example of Freddy, the main chimpanzee who adopted Oscar. While males can occasionally help a young in need (Boesch et al. 2010), the frequency of care given by Freddy shows a very particular connection (especially considering they are not related). It is not an unimportant epiphenomenon; the high, sudden frequency at which Freddy cares for Oscar gives information on the developing (he brings him into his life overnight) but very close connection (he takes care of him for months with a diligence rarely seen in males) with the orphan.

It is impossible for a researcher to have all the elements influencing the behaviour of an individual. However, it is possible to gather the ones they know and the ones they do not know to create a coherent and exhaustive outline.

We should first remind ourselves of the crucial spot held by the group in social species. This group is the stage for an incredibly large amount of micro-events, all of which may have an influence on an individual. In the last several years, ethologists have placed the behaviour of individuals into their hierarchical context – not just taking into account the fact that there are the dominant and the dominated but that individuals develop friendships, associate or protect each other. Age is also a factor that should be accounted for in species with a long lifespan, with the subtleties it sometimes involves (not all juveniles are juveniles in the same way, and the same goes for adults). The point is not to explain away an individual's behaviour, as their lack of maturity or unfinished mental development are fairly obvious, but rather to consider age as a modifying factor when dealing with other members of the species. Some behaviours from younger members of the group will not be addressed by others, or will be pacifically rebuffed when they would trigger a violent response if they were produced by fully-abled, virile males (De Waal and Robert 2002).

The group is also a fragile environment. It would be wrong to believe that there is no impact in an individual from events such as the perturbation of hierarchy, conflicts within the group (even if the subject has no part in it) or the death of an unrelated member of the group just because the subject did not have a direct connection to those events. A disease, a wound, a birth are some of the events that we would be remiss not to account for just because the subject is not involved. The group is a signifying super-entity that constantly emits signs – voluntarily or not – that influences its members. This factor is taken into account in different ways by ethologists. Primatologists that remain in contact with the same group for months, and sometimes years, have integrated this aspect in their research. Unfortunately, however, this taking into account grows weaker as the time spent observing is lower. Any modifier in lifestyle, or group dynamic or balance should be considered a modifier in the observed individual.

The individual themselves can be affected by a number of parameters that change the meaning of their semiotic acts. A particularly calm individual that suddenly becomes aggressive does not say the same thing if they are wounded, than if they are pregnant, or if their physical state has not changed. It is sensible to think the physical and mental state are closely linked: a subject that is currently vulnerable will be more easily affected by stressful events, the feeling of abandonment or aggressions (even minor or simulated ones), and similarly, they will not express their anger or jealousy in the same way as if they are fully abled, as they would be unable to defend themselves if their attitude were to lead to an open conflict with someone else. A particularly interesting point is the expression of emotions such as distress or sorrow: does an individual diminish their semiotic acts to not bring too much attention on themselves when in a position of weakness, or do they amplify them to attract the protection, comfort or support from the rest of the group?

While these contextual variables are easily accessible to researcher, we estimate that there are two that are just as important but much more difficult to find out about and therefore to take into account.

The first is what I can call the subject's "previous experiences". I return to the relevance theory here, and it tells us that everything that a subject emits, receives and interprets is connected to their mental build. This of course includes what the subject is biologically speaking, and what they are trying to do at that exact moment, but also how they were built over time, how their brain has organised itself to deal with its environment, connect to other members of the group and sometimes, survive traumas. In the case of animals in natural reserves, or individuals coming from sanctuaries, rescue or reintroduction programs, researchers may be aware of the subject's past and their emotional way of operation as well as how they express it, how it can be interpreted in light of their years of captivity, how close they are to humans, or trauma such as the death of their original group, an early loss of their parents, potential wounds or even in some cases a long-term form of post-traumatic stress disorder.

The second variable is what I could call the personality of the individual. This concept has long been kept away from the subject of animals but is well known by veterinarians, animal carers in zoos or sanctuaries, as well as researchers that study animal in the wild. This subject is not that one, and as observations continue, it becomes easier to predict completely opposite behaviours from two individuals with different personalities in the same situation. This variable, although still neglected, is possible to take into account but sometimes requires years and years of observation, which is not always possible.

4.2 Semiotic Concepts

In the second part of this chapter, I will introduce the semiotic concepts that make up my methodological model with the tools I discussed earlier. These concepts are ways to read and interpret data obtained from using those tools.

I will begin with the triple concept of Intentional-Conscious-Unconscious to show the mode of sign emission by the subject. This concept is directly linked to the notion of conscience mentioned in Chap. 3 as well as the voluntary aspect of "saying" as explained in Chap. 2.

This triple concept is also essential to guide the analysis of the semiosis and to offer a coherent interpretation with the next semiotic concepts.

I will also briefly discuss the phatic function of language as told by Jakobson (2003). This function, which may seem anecdotic next to the importance and complexity of the other functions of language, is actually a powerful interpretation tool that allows a coherent reading of certain voluntary signs, especially two categories: signs that are not very intense but frequent with no apparent use on one hand, and intense, rare signs on a different scale (for example, very frequent signs in a specific individual that are completely absent from the rest of their group) that carry a strong meaning.

Following this, I will present a few concepts from the semiotic theories of Umberto Eco that are interesting to work with in animal studies. It goes without saying that not all the concepts from Eco's semiotics are adapted to this type of work (far from it), so I will only mention the concepts I need to add to my interpretation, explain why I chose these ones and what they bring to the understanding of animal semiosis.

Lastly, I want to conclude this chapter on the particular and potentially problematic case of anthropomorphism. I have previous explained (Chaps. 1 and 2) how this concept is controversial and closely linked to various problems in animal studies.

However, I found it used as an interpretation tool in several authors during our research, and I believe that, if it is used with discernment, it can help with providing a definite answer in cases where the interpretation of semiosis is faced with an insolvable problem. For example, this will be the case when our interpretation will take us to the "black box" of the animal mind, a "final frontier" that I will discuss in detail in Chap. 7.

4.2.1 Intentional, Conscious, Unconscious

There exists a number of voluntary signs that reveal complex concepts that can be used to understand communication, cognitive abilities or emotions. This is the case of premeditation.

For a long time, it was denied that animals could project themselves into the future, understand the consequences of their actions, or simply to be fully conscious of their actions. Over time, proof of coherent cognition was gathered and listed (for example, the case of crows tested on by ethologists that proved themselves capable of following a succession many complex, linear steps to solve a problem, or this orang-utan female who slowly but surely removed the screws on the bars of her cage so that nobody would notice (Christen 2011).

I can mention two very distinct behaviours as examples, with the first being part of a category I could call "protective behaviours". It is the act of an individual adopting a number of different behaviours to protect themselves from a negative emotion they anticipate (justifiably or not). These behaviours are an indication of the correct understanding of causality between events from the subject and their ability to look to the future as well as anticipating their own mental states. This projection allows them to decide that they want to avoid some of them (sadness or the feeling of abandonment, for example), sometimes at the cost of a large amount of effort.

A striking example of this type of behaviour comes from dolphins, who will take care of a sick or wounded calf by carrying it, feeding it, and defending it to try and keep it alive. The calf may not survive regardless, and in this case, the group rapidly goes back to being calm, which is not what happens when a calf dies suddenly and brutally, which causes a large amount of agitation and many signs of emotional distress (King 2013). Dr. Joan Gonzalvo[3] believes that the school of dolphins tries to avoid death as much as possible but accepts it when it occurs and when it was prepared for, which is not the case when a violent death happens as it catches the individuals off-guard and leaves them powerless, without having allowed them to prepare themselves for it.

A more modest, but interesting nonetheless, example comes from adult suricates when they remove the venomous stinger from scorpions to let their young safely train to hunt (Thornton and McAuliffe 2006). While it is not necessarily a question of emotion, we can see that this type of behaviour indicates the existence of a theory of mind (putting oneself in the place of the inexperienced young) and can be seen as a form of empathy (trying to prevent a very painful wound for the young).

The second form of voluntary signs that we want to discuss here is the signs that aim to hide other signs. We must look at subjects that are aware, not just of the signs they emit (sometimes involuntarily) but also of the way that other individuals (not necessarily of the same species) perceive these signs, to try and influence that perception. The simple act of concealing an emotion by controlling their manifestations (mainly those that can be violent, such as the manifestations of jealousy or anger) can be mentioned as part of this category.

But I can also include in this category the voluntary signs that are part of particularly complex behaviours, such as manipulation or even lying. These cases have not been studied very much, most likely due to the preconceived notion that what is "wrong" and its manifestations (such as lies, manipulation, deceit, etc.) were intimately linked to the human species (and maybe even to just the modern, social human, as the myth of the "noble savage" is a hard one to get rid of), and could not exist in other species. The first contradictions of this came once again from primatologists, most notably a case reported by Penny Patterson. Koko once broke the sink in her living space (Schroeder et al. 1990) and tried to put the blame on one of her carers when Patterson asked her what happened, by signing "Kate", "sink" and "bad". We can only speculate on the reasons that led Koko to lie (additionally by accusing someone that had done her no wrong) but the emotional reasons (a fear of being reprimanded, or of Patterson's anger, a feeling of shame, etc.) are without a doubt the most probable cause and the most likely to push someone to twist reality.

The case of manipulation, especially the manipulation of emotions, shows the same blueprint. De Waal mentions individuals exaggerating their reactions (and therefore the perception that others have of their feelings) to cause the members of their group or nearby humans to pay attention to them (it is mostly a case of exaggerating pain to incite their peers to help them fight off an attacker, or anger and offense in situations of jealousy, to make the other individual care for them and offer gifs for reconciliation).

---

[3] Member of the board of directors of the Tethys Research Institute, director of the Ionian Dolphin Project and scientist working with Earthwatch.

The opposite of the production of voluntary signs and the possibility of consciously manipulating them is signs being emitted in a complete involuntary but still conscious manner. Even though they are opposites, these two forms of signs can perfectly work together as well as change their interpretation for semioticians or biologists.

Voluntarily emitted signs coexisting with contradictory, involuntary signs are a special and very telling combination of signs. I discussed earlier the idea of voluntarily concealing emotional manifestations, but that is not necessarily an easy thing to do, as emotions follow complex psychological, physical and physiological processes and not all manifestations of a strong emotion can always be concealed. A human example would be a speaker that is highly trained to control the tremor in their voice or their speech rate (not an easy task for a novice) but has no influence on the dryness of their mouth, a slight quiver or their heart racing – and a particularly careful observer can notice them.

I believe that the more an individual is subjected to complex processes, the more important the manifestations are (not necessarily visible, but important nonetheless) and the more it impacts them on a mental and physical level. The more they occupy the mental world and the physical state of an individual, the more the manifestations of these processes are made up of very diverse and varied micro-signs, and the harder they are to control completely.

Why try to hide signs, and why are these signs that cannot be hidden? The second question is the easiest to answer, as many specialists have studied the subject, mostly in humans, but a number of conclusions made about our species can be applicable to others.

It is the case, for example, of tears of sorrow. Studies suggest that humans cry and struggle to hide their tears to convince others of their emotional sincerity and the degree of distress, which means they require some attention, comfort, and maybe even some help (Bohler 2007). At this time, it has yet to be proven that other species (even amongst great apes) are capable of tears of sorrow, it is sensible to think that other species are afflicted by the same difficulty in hiding signs of distress for similar reasons.

Semioticians should therefore pay close attention to these involuntary signs, even when they coexist with *seemingly* contradictory voluntary signs, because they are the ones that tell us the most about the real mental state of the subject despite their attempt to conceal it.

This leads us back to my first question – why want to conceal exterior signs of processes? Regarding emotions, it should be remembered that emotional manifestations are signs for members of one's own species first, and for any individual getting in contact with the subject (predator, hostile or foreign subject, etc.) second. They mostly exist because they are necessary for the protection, security, and managing of the social fabric of the individual.

However, as an individual develops strong, complex social ties, the effects of those manifestations on their social fabric also become more important, varied, and depending on their social goals, can help or hinder them. Not knowing how to control their anger or letting their jealousy appear is risking the creation of hostilities with other members of the group (De Waal and Robert 2002). Fully displaying dread or sorrow is letting others see a state of relative weakness that they can exploit.

The cohabitation of voluntary signs and conscious but involuntary and contradictory signs must be seen as a strong indicator of the existence of a theory of mind in an individual, the ability to hand to others thoughts and intentions depending on what the individual lets them see, as well as the will to manipulate this perception to get an advantage.

It can also be seen as an awareness (at least a rudimentary one) from the individual of the impact of their emotions on their own state or their existence in general (De Waal 2013). Clues of this type of conscience can be found (even if, in this case, it is an awareness of others) in the individuals close to someone experiencing pathological grief. The members of the groups of dolphins or chimpanzees never try to take away the body of the young from its mother, nor do they try to move her away from it. They accompany her until she decides to part with the body, and only then will they stop paying attention to it. It is possible to see, in cases such as these, an elementary form of understanding of the strong emotional state as disrupting the normal behaviour of the individual.

As mentioned earlier, it is not always easy to know when an individual is conscious of the signs they emit and at which degree, especially when they are emotional signs. Voluntary signs can be handled by the individual as unconscious signs, that is to say there is no specific reaction to those signs – they do not try to hide them, either because they stand by them or because they do not realise they exist.

However, the circumstances can give us more information and help categorise the signs we are witnessing. For example, when faced with an individual emitting intense signs of emotional distress, such as Washoe when learning about the death of Sequoyah or the vast number of pets that let themselves die when their owner passes,[4] it is epistemically costly to say it is possible these signs are voluntary. Depression, even in humans, is a state that produces behaviour that the subject is often unaware of and will forget later on. And that is where the danger lies: without external assistance, the subject is incapable of noticing the reduction of their food intake, their withdrawal into themselves, or the excessive use of various psychotropic drugs. The need for humans to require medical, and often chemical, assistance leads us to think that signs of emotional distress are more unconscious manifestations of a paralysing state of mind rather than voluntarily emitted signs to ask for help, for example.

So I want to make a methodological choice here, which is that we choose to consider by default that signs are unconscious if they are not an indicator of inten-

[4] One of the oldest examples we found in literature comes from Pliny The Elder's *Natural History*, book VIII: "Jason, the Lycian, having been slain, his dog refused to take food, and died of famine". However, there are very few contemporary ethological studies on the subject (while there are many studies on the grief of humans due to the loss of an animal) and examples must be searched for amongst veterinarians and behaviour scientists (Béata 2013, p. 291).

tion (even an unclear one) from the subject or if said subject does not give them some form of attention (as is the case with involuntary but conscious signs, that they may try to control). Of course, this label is never definitive and any clue that may lead us to think that this or that semiotic manifestation is actually conscious and voluntary must be examined.

This is not to fall back into a behaviourist mindset that would think that animals never, or almost never, have any awareness of their actions, communication or behaviour, and that they just act perfectly instinctively and unconsciously all the time. On the contrary – I believe that some animal categories (Bekoff 2009), of which humans are a great and undeniable representative (Bohler 2007), possess a continuous (throughout the day), permanent (throughout their life) emotional life that is abundant enough to regularly manifest itself through perceptible semiotic acts, without the subject always noticing them.

These acts do not always teach us much about the way individuals see their own emotions or even how they feel them. But they do tell us about the state of a subject at a specific moment (Washoe's depression, the nervousness of elephants as lionesses approach, the manifestations of happiness in adolescent chimpanzees when they are allowed to hold a baby in their arms, etc.) as well as the way their emotional life functions. The fact that elephants touch and are generally interested by the skulls of their dead family members tells us a lot about their long-term memory and the depth of their social ties, but the fact that they do it completely silently and so delicately (for a mammal that weighs six and a half tonnes on average) says a lot more about their feelings about these skulls. What are those feelings? We do not have a clue and we may never find out. But what we do know is that they are important enough in their mind to make them be extremely cautious, and strong enough in their mental world to force them to be silent while they handle the skulls.

However, not all species are affected by these "emotional leaks", and it is important to try and establish why they are not. First, it goes without saying that a permanent emotional life requires elaborate cognitive skills that are not present in all branches on the animal kingdom. While this does not mean these branches cannot have temporary emotional capabilities, simple feelings or sensations, it is not an "emotional life" in the way we are describing it.

Second, any information leak is a potential danger in the wild, and the species that can "allow themselves" to have such an abundance of semiotic acts are by definition species in which individuals evolve in a relatively secure environment: a group. Solitary species should not be considered as unable to have a rich emotional life, but it seems sensible that theirs would be more discreet. It is also possible that these semiotic manifestations, as unconscious as they may be, still play an important role in the communication and cohesion inside a social group (as we saw with emotional distress, what is invisible for a subject can be visible to others around them) and therefore are an evolutionary survival tool that would not need to exist in a solitary species.

4.2.2 *Jakobson's Functions of Language*

Jakobson's functions of language (2003) are very specific and some are probably exclusively human. But we can use one of them, the phatic function, due to the fact that it shows the point of attempts to communicate rather than the message itself. This is in direct correlation with the voluntary aspect of saying as we explained it in the previous chapters.

The voluntary aspect of communication indeed is important to understand the effort that an individual can put in their language performance. I briefly mentioned earlier the case of Koshik (Stoeger et al. 2012) which I will now explain in details. Koshik is a male elephant capable of articulating words in Korean. While it seems the elephant is unable to use the words he pronounces correctly, his greatest performance is on the phonological level. The elephant's upper lip is bound to its nasal bone and, as a result, cannot be used to perform an articulated pronunciation. Koshik has resorted to an extremely complicated technique that consists of placing his trunk in his mouth so it can play the role of his upper lip. Using this, the elephant manages to articulate a few discernible words. The researchers that have studied his case believe he does not understand the meaning of the words at all, or at best very slightly. However, they stress that the technique itself is remarkable, and the words he pronounces are audible and identifiable.

While the performance is impressive, it raises an important question: why has this animal tried to achieve something so complex? The reason is simple: Since arriving to the zoo, Koshik has been alone and his social contacts are mainly with his carer, who spends a lot of time with him. His efforts have one goal in mind; making that individual from another species understand that he needs social interactions with them, and that he wants to communicate with them, even if his grasp on language is unclear and inaccurate – pet owners act similarly when they imperfectly mimic their cat or dog while they play with them.

This voluntary aspect of "saying" can therefore be entirely interpreted through the lens of the phatic function of language, that Jakobson says has the role of establishing, maintaining or verifying the contact between the emitter and receiver (Jakobson 2003, chap. 11). Granted, the performance by Koshik and others that have tried a similar approach (another famous case is Noc, a beluga whale raised in captivity who would change the frequency of his calls to have them match the voice of humans, without losing his ability to communicate with other members of his speeches and without it being attributed to any specific physiological condition (Ridgway 2001)) is not discourse proper, but the phatic function is present nonetheless: all of this is the result of the elephant trying to create and maintain contact with his carer, and maybe other humans that may be nearby.

I could sum up the utility of the concept of the phatic function in a semiotic model applied to animals in this way: the voluntary aspect of "saying" is close to the phatic function of language, which highlights the means (cognitive, temporal, energy…) through which the emitter tries to establish communication and/or through which the receiver tries to perceive the message. Depending on the contextual elements of interpretation, this function of language can tell a lot about the motivations of the individual, that is to say, why they are ready to go through so much effort, and in turn, the degree of importance that the source of this motivation has in their mental world.

4.2.3 Eco's Semiotic Theory

This voluntary aspect of "saying" had already been made explicit by Eco in 1987; the act of communicating exists because there is an urge, a will, a search for collaboration from the emitter to the receiver (Eco 1987). The idea I am displaying here is not new, but it had not yet been explored as a study approach for zoosemiotics.

However, the semiotic theories of Eco are incredibly vast and rich and it is obvious that not everything in them is relevant to animal studies. I will only introduce the concepts from these theories that can help us interpret properly the signs revealed by the other tools.

Because on top of having brought a significant amount of academic support (as explained in previous chapters) to it, Eco also worked closely with zoosemiotics, notably with Sebeok (Eco and Sebeok 1983). Together, they devised methodological perspectives that can be applied to any zoosemiotic research despite coming from all the challenges they met while working in semiotics. This is very much the case of the methods of research through induction or abduction described by Sebeok (Eco and Sebeok 1983 p. 8) "Deduction: Rule – Case – Result; Induction: Case – Result – Rule; Abduction: Rule – Result – Case" that him and Eco summarise and justify thusly:

> « I never guess » Sherlock Holmes in The Sign of Four – But we must conquer the truth by guessing, or not at all. (Eco and Sebeok 1983 p. 11)

Another concept, that was solely expressed by Eco, also seems important to underline. A sign system does not need to be incredibly complex to be interesting semiotically. Nature is efficient, what is kept is what is best for survival at the minimal amount of effort, and this rule includes our species: for example, we are capable of having an extremely complex and specific vocabulary, but people that are not involved in those circles know nothing about technical vocabulary, scientific jargon or medical terms, and that does not cause them any problems. It is not because most of us are incapable of learning those terms, but rather because we have no use or feel no need for it. Eco (1999 p. 97), who was already quoted in Chap. 2, pointed out an ironic facet of this idea of complexity that we have:

> Kant fait très certainement partie de ceux qui ont persuadé les philosophes de la légitimité qu'il peut y avoir à construire une phrase qui compte vingt-deux lignes dans l'édition de l'Académie sans un seul point.

Translation:

Kant is certainly one of these people that managed to convince philosophers that it is legitimate to build a sentence that is twenty-two lines long in the Academy edition without using a single dot.

I have one last concept linked to the notion of sign that we have to discuss, even if it may seem very specific, because it actually has a vast field of application: the case of the supersign.

This concept does not really come from the interest that Eco had in zoosemiotics. Rather, it comes from an entirely different context, his studies in medieval aesthetics (Eco 1997). But the demonstrations he makes in his book show without a doubt that it is a very adaptable, shape-shifting concept that is useful in all the different forms of semiotic analysis, including the one applied to animal studies.

What is a "supersign"? It is the combination of several signs that, put together, form a unit that is both coherent on its own and larger than the sum of its parts. To give a clear example, picture a work of art such as a painting, which, in itself, is a semiotic entity, but it is also composed of individual elements (motifs, shapes, characters, etc.) that all have their own semiotic value, and which sometimes have additional links to other elements, to overall end up with the "supersign" of the painting. For this reason, amongst others that were mentioned earlier, the subject of Art is somewhat of a special case, despite its strong emotional charge and the important cognitive abilities it seems to require, that we will not explore in this book.

This concept can be particularly useful when it comes to the handling of complex semiotic processes, such as modes of communication with multiple channels (such as dolphins that use sound and sonar simultaneously to communicate (Ryabov 2016)) or emotions. I think that, for example, maybe complex manifestations, especially rituals, are born out of a need for a response to an emotional event. And rituals are inherently supersigns. They are made up of gestures, attitudes, objects, sometimes in a specific order or with specific attributes (colour, shape, position).

It may occur that some signs have no particular meaning on their own and only take one on in the context where they play that role of fragment of a greater sign (staying with the example of grief, an individual collapsing, if not due to a pathological condition, is not a sign that has a specific meaning, it can simply be the result of exhaustion – but when this collapse occurs in the context of a mourning ritual, it suddenly takes on a strong connotation of sorrow and distress).

Therefore, the semiotic analysis of such a manifestation requires not only the analysis of its elements, but also the connections those elements have between them and the global meaning that derives from them.

The question of sign justification makes a resurgence with that last point: if an isolated element of the ritual can be justified in a perfectly rational manner (anointing the body in a mourning ritual so as to not have it attract predators, etc.), it is hard to do so with the supersign as a whole (the systematic existence of a ritual related to grieving is about a lot more than simply anointing the body (van Leeuwen et al. 2016)).

The supersign is also interesting in that it enables us to observe each of its components to give several sub-answers to each part of a question. Going back once more to the subject of mourning, that is a response to a situation: taking into account the death of an individual. But the ritual contains a number of smaller responses to smaller situations: burying the body is a response to the issue of handling the body, the wake is a response to dealing with loss, taking care of the loved ones of the deceased is a response to the emotional projection an individual does regarding distress and grief, returning later to the body and touching the skulls is a response to the question of remembrance and memory, and so on.

Behind each of these smaller elements of the supersign, there are cognitive processes that we need to see and understand (long-term memory, empathy, compassion, theory of mind, etc.), as well as understand the supersign as an extremely advanced and complex process in which all of its capabilities connect to each other to try and find a response to a complex problem. While Peirce says there is no thought without sign, to Eco, there is also a degree of complexity in thought at which the single, autonomous sign no longer suffices. Mentioning the concept of supersign makes us acknowledge this degree and how to handle and understand it.

4.2.4 About the Case of Anthropomorphism

It is reasonable to question the idea of using the concept of anthropomorphism as a means of interpretation in a scientific book. The very history of the study of animal behaviour is rife with errors in judgement that can be directly attributed to anthropomorphism. Far from being an indicator of sentimentality or an absurd conclusion, anthropomorphism comes from us always trying to understand foreign things by relating them to ourselves, which can lead us to attribute thoughts and ways of functioning similar to our own to others.

However, the main problem is not really that we look for what is similar to us in others, but rather that we have a tendency to believe that *only* what is similar to us can be a sign of advanced cognition, of emotion, of complex communication. But searching for similarities in others should not deny the fact that they are a specific being, distinct from us through their abilities and evolutionary history. Which Martinelli explains thusly in the case of language:

> [...] The capabilities (cognitive and practical) to acquire language are not inaccessible to other species, however human beings have been, and are, the species that appears to be the most interested in pursuing it, and – consequently– the best disposed to do so. (Martinelli 2006 p. 475)

This is where relevance theory will fulfil its first duty as a "sorting tool". By using the accumulated knowledge we have on a species, we can tell whether this or that concept is relevant to their mental world and their functioning, and as a result, whether an explanation based on those concepts is possible.

To explain my remarks so far, I will use a famous case of an error due to anthropomorphism that has now become a textbook case: Hans the horse, also known as "Hans the Clever".

Here is a brief summary: Hans was a horse raised by a German owner at the beginning of the twentieth century, and seemed capable of feats of intelligence such as solving simple operations, spelling words or solving problems of musical harmony, by using as a code the striking of one of his hoofs on the ground. After several years of study, researchers that had been intrigued by such capabilities ended up concluding that Hans did not, in fact, know how to count or read, but that he was just providing answers based on almost imperceptible signs that his interlocutors were unknowingly sending him (Rosenthal 1970).

I will know look at how relevance theory can help us on this subject, and how it can allow us to handle cases such as this one.

First – do the capabilities that were attributed to Hans the Clever have any chance of existing without his species, and are they permanent capabilities?

Horses, in their natural state, are herbivores and preys. While there have been questions from researchers in the past few years about the role of our own predation in the development of our calculation faculty, this faculty only has a very limited use for animals that do not hunt. Some species are the exception, however, such as elephants who, despite being herbivores, seem to show some numbering capabilities (their behaviour changes completely whether they are dealing with one lioness, who is not a real threat, or a small pride of lionesses, usually three, which is a threat to a younger elephant or a weakened member of the group.

How do they use those capabilities, and could horses have the same use for them? When faced with predators threatening them of their young (especially the latter for elephants), the two species have very different reactions. Horses flee, preferably in compact groups (Béata 2013, p. 37). Elephants choose a defensive position, in a circle, (Christen 2011, p. 93), with the young at the centre (this position is used whether they are facing predators or a hostile family of elephants). In some cases, the matriarch, sometimes accompanied by several females, will launch a charge towards the predator (Ameisen 2013 pp. 56–57). But this choice is extremely risky is the threat is composed of several individuals, and becomes much more dangerous as the number grows. It is therefore logical to think that numbering capabilities (even rudimentary ones: one, a few, many) are very useful to elephants for choosing their response to predators, which is not the case for horses.

The existence of this capability is very unlikely in this species, and the same goes for the others that were attributed to Hans by his owner. It is not that the horse is not *capable* of reading, counting or understanding music; it is simply that those skills are not *useful* to him, and as a result, the chances of him and his species developing them during its evolution are very slim.

However, what can I deduce with the relevance of the conclusions that were reached regarding this phenomenon by the scientists at the time? They seem perfectly coherent. A horse like Hans is an animal domesticated by a human, living with him, by him and sometimes thanks to him. His survival (at worst) and his well-being (at best) entirely depend on his ability to create social ties with members of the human species and understanding them as well as possible. In his mental world, the signs, expressions and emotions of humans are extremely relevant elements, and it makes perfect sense that he ended up developing tools to interpret them as best he could as he grew up.

A closer look shows that the problem here is not really anthropomorphism. The mistake that we must absolutely avoid is considering the observed behaviour (here, the advanced reading of signs in human faces) as not intelligent, or at least not as intelligent as knowing how to count. The signals emitted by his owner were almost entirely invisible to the human eye. Knowing how to notice then and how to interpret them show a great emotional capability and the probably existence of a theory of mind, both of which are very advanced cognitive aspects. As a reminder, Christen (2011, p. 167) defines theory of mind as such:

> De quoi s'agit-il ? Tout simplement de l'aptitude à se mettre mentalement à la place d'un autre, à lui attribuer un esprit. De là découle une possibilité d'emboîtement des états mentaux, [...] propice à assurer la complexité d'une vie sociale riche
>
> Translation:
> What is it? Simply the ability to put oneself in the place of someone else, to give a mind. From there comes a possibility of an interlocking of mental states, [...] suitable to guarantee the complexity that comes with a rich social life

Which naturally leads me to conclude that, here, it was not the (possibly anthropomorphic) perspective we had on the animal that was the issue, but rather, as developed in Chap. 2, the perspective we have of ourselves as an ideal that needs to be reached that is the cause of error. Other species do not develop skills to try and be similar to us, but because they are useful to them. To look for similarities in another is sensible (more or less, depending on the evolutionary proximity), to only pay attention to this is deleterious.

Some, such as Marc Bekoff (2009) for example, see anthropomorphism as far from a mistake in judgement but rather a precise cognitive tool to know others. He is strongly supported in the preface of his book by Jane Goodall; therefore, I believe that his school of thought, which is far from being a minority, has a certain foundation amongst the scientific community.

I agree with them on this point in this book, and I also think that anthropomorphism, when filtered through the relevance theory, can be a concept that helps with interpreting animal semiosis, as soon as we are able to see our point of view as what it truly is: the perspective of an animal that is part of the human species, and therefore limited by a number of restrictions that apply to it. The question of whether we should say a dog is jealous, or an elephant is sad, due to the pretext that we do not know what they are feeling exactly, is a non-issue. Our mind does not let us have the certainty that others, even other humans, feel the same way we feel. We can only tell ourselves (to borrow a phrase from quantum physics), "everything acts as if" our peers feel the same sadness, joy, anger or fear. By admitting that it is, and will remain, impossible to know more than that, it seems relevant to treat animals the same way when the signs allow it (as well as cerebral imaging, once it finds similar neural reactions in various species): "everything acts as if" the emotion an animal is feeling is similar to the one we would feel in a similar context. This is not the exact truth, but it is as close as we can get to truth without ever being able to reach it.

References

Ameisen, J.-C. (2013). *Sur les épaules de Darwin : Les battements du temps*. Mayenne: Les liens qui libèrent.
Barbieri, M. (Ed.). (2006). *Introduction to biosemiotics: The new biological synthesis*. Dordrecht: Springer.
Béata, C. (Ed.). (2005). *Zoopsychiatrie : La communication*. Marseille: Solal Editeurs.
Béata, C. (2013). *Au risque d'aimer*. Paris: Odile Jacob.
Bekoff, M. (2009). *Les émotions des animaux*. Paris: Éditions Payot & Rivages.
Boesch, C. (2007). What makes us human (*Homo sapiens*)? The challenge of cognitive cross-species comparison. *Journal of Comparative Psychology, 121*, 227–240.
Boesch, C., Bolé, C., Eckhardt, N., & Boesch, H. (2010). Altruism in Forest chimpanzees: The case of adoption. *PLoS One, 5*(1), e8901.
Bohler, S. (2007). *La Chimie de nos émotions*. Genève: Aubanel.
Christen, Y. (2011). *L'Animal est-il une personne ?* Barcelone: Flammarion.
Cole, P. (Ed.). (1982). *Speech acts*. New York: Elsevier.
de Saussure, F. (2002). *Ecrits de linguistique générale*. Paris: Gallimard.
De Waal, F. (2013). *Le Bonobo, Dieu et nous*. Lonrai: Les Liens qui Libèrent.
De Waal, F., & Robert, M. (2002). *De la réconciliation chez les primates*. Paris: Flammarion.
Delahaye, P. (2013). *Application de la zoosémiotique à l'éthologie : peut-on parler de culture animale ?* Paris: V Descartes.
Deledalle, G. (1971). *Le Pragmatisme*. Paris/Montréal: Bordas.
Eco, U. (1987). Notes sur la sémiotique de la réception. *Actes Sémiotiques Documents, IX*(81).
Eco, U. (1988). *Sémiotique et philosophie du langage*. Paris: Presses universitaires de France.
Eco, U. (1997). *Art et beauté dans l'esthétique médiévale*. Paris: Grasset.
Eco, U. (1999). *Kant et l'ornithorynque*. Paris: Grasset.
Eco, U., & Sebeok, T. A. (Eds.). (1983). *The sign of three: Dupin, Holmes, Peirce*. Bloomington: Indiana university press.
Favareau, D. (2006). The evolutionary history of biosemiotics. In M. Barbieri (Ed.), *Introduction to biosemiotics: The new biological synthesis* (p. 30). Dordrecht: Springer.
Favareau, D., Cobley, P., & Kull, K. (Eds.). (2012). *A more developed sign: Interpreting the work of Jesper Hoffmeyer*. Tartu University Press: Tartu.
Fleming, S. (2009). Les corbeaux ont-ils une cervelle d'oiseau ?. Arte France.
Fodor, J. (1983). *The modularity of mind: An essay on faculty psychology*. Cambridge: MIT Press.
Fouts, R., & Mills, S. T. (1998). *L'école des chimpanzés: ce que les chimpanzés nous apprennent sur l'humanité*. Paris: J. C. Lattès.
Grassé, P.-P. (Ed.). (1955). *Traité de zoologie, anatomie, systématique, biologie, Tome XVII. Mammifères, les ordres : anatomie, éthologie, systématique. 1er fascicule*. Paris: Masson.
Grice, P. (1982). Logic and conversation. In P. Cole (Ed.), *Speech acts* (pp. 41–58). New York: Elsevier.
Hoffmeyer, J. (2008). *Biosemiotics. An examination into the signs of life and the life of signs*. Scranton: University of Scranton Press.
Jakobson, R. (2003). *Essais de linguistique générale, 1: Les fondations du langage*. Paris: Éditions de Minuit.
King, B. J. (2013). When Animals Mourn. *Scientific American, 2013*, 62–67.
Kull, K. (2012). Advancements in biosemiotics: Where we are now in discovering the basic mechanisms of meaning-making. In S. Rattasepp & T. Bennett (Eds.), *Gatherings in biosemiotics* (p. 18). Tartu: University of Tartu Press.
Martinelli, D. (2006). In M. Barbieri (Ed.), *Introduction to biosemiotics: The new biological synthesis* (p. 475). Dordrecht: Springer.
Massal, N., & Lachapèle, D. (2014). *Mon chat est jaloux : mieux le comprendre et le rassurer au quotidien*. Paris: Eyrolles.
Moss, C. (1989). *La Longue Marche des éléphants*. Paris: Robert Laffont.

Nissani, M. (2006). Do Asian elephants (*Elephasmaximus*) apply causal reasoning to tool-use tasks? *Journal of Experimental Psychology: Animal Behavior Processes, 32*(1), 91–96.
Patterson, F. (1986). *The mind of gorilla: Conversation and conservation. Primates: The road to self-sustaining populations*. New York: Springer.
Peirce, C. S. (1987). *Textes fondamentaux de sémiotique*. Paris: Méridiens-Klincksieck.
Peirce, C. S. (1992). *Ecrits sur le signe*. Paris: Éditions du Seuil.
Pliny the Elder (Gaius Plinius Secundus), *Naturalis Historia*, Vol. III, Book 8.
Pruetz, J. D. (2011). Targeted helping by a wild adolescent male chimpanzee (Pan troglodytes verus): Evidence for empathy ? *Journal of Ethology, 29*, 365–368.
Rattasepp, S., & Bennett, T. (Eds.). (2012). *Gatherings in biosemiotics*. Tartu: University of Tartu Press.
Ridgway, S. (2001). Hearing and whistling in the deep sea: Depth influences whistle spectra but does not attenuate hearing by white whales. *Journal of Experimental Biology, 204*.
Rosenthal, R. (1970). Le préjugé du maître et l'apprentissage de l'élève. *Revue française de pédagogie, 13*, 39.
Ryabov, V. A. (2016). The study of acoustic signals and the supposed spoken language of the dolphin. *St. Petersburg Polytechnical University Journal: Physics and Mathematics, 2*(3), 231–239.
Schroeder, B., Patterson, F., Fouts, R. et Pribram, C. (1990). Koko le gorille qui parle. Warner home video France.
Sperber, D., & Wilson, D. (1989). *La pertinence: Communication et cognition*. Paris, les Éditions de Minuit.
Stoeger, A. S., Mietchen, D., Oh, S., de Silva, S., Herbst, C. T., Kwon, S., & Fitch, T. (2012). An Asian elephant imitates human speech. *Current Biology, 22*, 2144–2148.
Swan, L. (2012). Scholarship. In D. Favareau, P. Cobley, & K. Kull (Eds.), *A more developed sign: Interpreting the work of Jesper Hoffmeyer* (p. 232). Tartu: Tartu University Press.
Thornton, A., & McAuliffe, K. (2006). Teaching in Wild Meerkats. *Science, 313*(5784), 227–229.
van Leeuwen, E. J. C., Mulenga, I. C., Bodamer, M. D., & Cronin, K. (2016). Chimpanzees' responses to the dead body of a 9-year-old group member. *American Journal of Primatology, 78*(9), 914–922.

Chapter 5
Intertheoricity: How to Build Bigger Models

Abstract Creation sub-question n°1: How does intertheoricity allow us to build bigger and more efficient models in animal studies?

This chapter will introduce the concept of intertheoricity. Thus, it will return to the current academic positions regarding interdisciplinarity, to the difficulties brought by this operating mode and the way Guillaume's theory of intertheoricity suggests solving these issues. Following this, the chapter will analyse our model of animal semiotic study, by showing how it is enabled by intertheoricity: thanks to the development of a joint methodology for language sciences and life sciences, thanks to the harmonization (and creation when needed) of definitions between fields, and thanks to opening a dialogue between different concepts across fields. Finally, this chapter will explain why such models are necessary today, whether it is to study more complex subjects, compensate for the over-specialization of researchers, or optimize the impact of research.

Keywords Intertheoricity · Interdisciplinarity · Linguistics · Biology · Theoretical models

In this chapter, I go fully into the "creation" part of my methodological model, and I will endeavour to answer the first question of this sub-category in my introduction: How does intertheoricity help us build bigger, more efficient models in animal studies?

While it has become more and more common nowadays to hear about interdisciplinary works, it is clear that the term "intertheoricity" is still almost never mentioned. As far as we know, in English, this term was almost translated specifically for this book (the author has used it several times in communications, but never in published articles). It can be found in French publications (and perhaps their English translations) by semiotician Astrid Guillaume (2014). While the idea is not necessarily new, she was the first to give a name to this methodological concept.

As I am aware that it is probably little known by the public, especially the non-French-speakers, I will begin this chapter by presenting what this concept of intertheoricity really is, when it is used, why it has been a necessary element in certain

P. Delahaye, *A Semiotic Methodology for Animal Studies*, Biosemiotics 19,
https://doi.org/10.1007/978-3-030-28813-6_5

academic branches or certain types of works, how Guillaume explains it and what the term really covers in the theory of its concept.

Following that, I will show how intertheoricity is a methodological tool that allows for building bigger study models, especially in the fields that interest us, a particularly important point in a work such as the one presented in this book, which aims to study a subject from one academic field while using tools that traditionally belong to another. I will go into more detail about the way intertheoricity enables setting up a methodology in common for the involved disciplines, how the definitions turn out to be, once again, a complex source of confusion, and how to solve this problem, most notably through harmonisation, and finally, how to have these concepts "communicate" with each other even when they belong to different academic fields, without losing or twisting their specific meaning.

I will end the chapter by answering the question many could be asking: why would we need bigger, vaster models than the ones we can easily have today? I will answer by reviewing three scenarios that I deem representative of the vast majority of relevant cases. The first will be the case in which we want to study more complex subjects, in the sense that they require more specific aspects that belong to different fields and are difficult to study in a "monolithic" manner with just one disciplinary field. The second case is studies that run into the issue of overly specialised researchers and require larger study models, usually because they are subjects that have not been studied very much. The last case will be studies that aim for a bigger impact (for example, by giving answers that can contribute to several disciplines) while also taking less time, energy and resources than putting together a study for each separate field.

5.1 What Is Intertheoricity

The term "intertheoricity" is a direct translation of the French term "interthéoricité" as it is used in the works of Astrid Guillaume. While it may seem to be fairly transparent and speaking for itself, I will start by detailing its bases and explain the role it has in the creation of our theoretical model.

First, I will talk about the current academic position on the subject of interdisciplinarity, as the concept of intertheoricity is closely linked to it. This aims to be a general review, but it will focus on the French academic position more specifically, as it is the context in which this concept was created. This review will also be the opportunity to see the positions taken by different disciplines and fields of research on the subject of interdisciplinarity, and therefore to see how intertheoricity can be more or less necessary, expected or neglected in those fields.

I will then take a closer look at the main current challenges and flaws of interdisciplinarity across all fields (while paying particular attention to the case that interests us, the semiotic methodology in animal studies). I will examine why these flaws are problematic for studies that need a solid method and how intertheoricity can help solve these problems.

Lastly, I will take the time at the end of this section to present the theory and work of Guillaume in more details. This will have two main goals. The first is to make clear the ins and out of this theory and show how my model is built on them. The second is to avoid some confusion in regards to attributing to this author the creation of this semiotic theory.

5.1.1 Academic Position About Interdisciplinarity

The current, most common academic position on interdisciplinary is relative ambiguous and unclear. While it is more and more encouraged on an official level, most notably towards national research organisations and research-focused bodies in Europe, this official spotlight does not really translate into actual use on the field.

It is possible that the French case is an exception once again. Other countries do not hesitate to encourage these interdisciplinary endeavours, a recent example mixing philosophy, education, engineering and empathy studies (Hess et al. 2017) with convincing results. But in the academic tradition we stem from (and consequently, that my model stems from as well), interdisciplinarity is more of a vain wish than a tangible academic reality.

As I have already discussed this particularity in Chap. 2, I will not return to it, but one point needed to be clarified. For the most part, researchers that tend to confine themselves into firmly monodisciplinary models are not wrong: those are models that have proven themselves in the past, and that are well-known amongst their fields, sometimes even internationally, which can involve some reworking to fit another academic culture. These models are overall great at saving time and energy, and are functional for most of the cases the researchers will be dealing with. To this, I add another particularity I believe to be symptomatic in the French research community: most of the projects presented as interdisciplinary often are not and do not need to be interdisciplinary works. For these projects, the choice of interdisciplinarity is often a political move rather than a scientific one, and inconveniences the researchers more than it helps them, resulting in a low credibility for the concept.

This still leaves the question of, what happens in those other cases in which an interdisciplinary approach is necessary to the project. There rarely is a methodological consensus on how to make the different disciplines work together; each team, each laboratory does that they consider to be the best suited option on their own, which does not lead to any real thought on the subject. There is currently almost no study on what an interdisciplinary methodology should or should not be. This is how Astrid Guillaume sums us this partial "methodological void":

> *On parle de plus en plus de l'intertextualité, d'intermédialité et d'interartialité ; l'interthéoricité, en revanche, est très peu évoquée et pratiquée dans les sciences humaines et sociales (SHS), elle ouvre pourtant la voie à une nouvelle manière de comprendre les théories d'hier et de créer celles de demain.* (Guillaume 2014 p. 17)

Translation:

There is more and more talk about intertextuality, intermediality and interartiality; intertheoricity, however, is seldom mentioned and used in human and social sciences (HSS), despite opening a door to a new way of understanding the theories of yesterday and creating those of tomorrow.

To Guillaume, this problem is actually mostly rooted into human and social sciences, which we will discuss in more detail next. We only partially agree with this analysis, as any problem that would exist within human sciences would compromise a possible collaboration amongst them, especially life sciences. Therefore, no matter where the problem starts, it quickly becomes a problem on a general level.

In this way, intertheoricity can be seen as an attempt to solve a visible problem in current academic positions: many projects need an interdisciplinary approach, but the way to do it, the methodology required to see these projects through is not really thought through. Intertheoricity is a concept that aims to acknowledge and accept the legitimate difficulty that disciplines have in working together, to understand the crux of these problems and provide methodological solutions.

5.1.2 Difficulties and Flaws of Interdisciplinarity

One of the first problems that any interdisciplinary work runs into is that not all disciplines can easily evolve or be hybridisable. While life sciences, by the very nature of their shared inherent theory, can evolve, are modifiable and are regularly modified (the publication of new articles often shakes up models that were thought to be cemented, as it has happened recently (Langer et al. 2017) with the classification of families and groups of dinosaurs), this is not the case with language sciences.

Guillaume (2013, 2014) therefore has some strong words about the theories of human sciences, which she describes as much more unwilling to change than other academic disciplines:

> *[Le fait qu'il est possible de faire évoluer une théorie en fonction des besoins expérimentaux] est nettement moins vrai dans les SHS, où les théories et les théoriciens sont beaucoup plus rares que dans les autres champs du savoir ; les contextes auxquels ces rares théories s'appliquent (langue, littérature, philosophie, société, psychologie) ne permettent pas des vérifications aussi chirurgicales que dans les sciences médicales, ou des applications aussi libres et libérées que dans les arts. S'il y a bien un domaine qui évolue lentement et assez peu en matière de théorisation pure, c'est bien celui des SHS : pour faire court et simple, les psychologues sont toujours à cette heure soit freudiens, lacaniens ou jungiens, les sociologues pro Bourdieu ou anti Bourdieu, les linguistes pro Chomsky ou anti Chomsky, créant depuis des décennies des écoles de disciples qui s'opposent avec de vraies querelles de clochers, ce qui ne fait pas pour autant avancer la recherche, vu que ces querelles se cantonnent à préserver la mémoire théorique d'un ancien. Cette attitude tend à laisser penser que les théories en SHS ne sont ni élastiques ni hybrides. Elles ont, certes, toute une plasticité qui leur est propre mais elles ne sont pas toutes élastiques et hybrides.* (Guillaume 2014 p. 12)

Translation:

[The fact that it is possible to have a theory evolve depending on the experimental needs] is much less true in the HSS, where theories, and theorists, are much rarer than in other fields; the contexts in which these rare theories are applied (language, literature, philosophy, society, psychology) do not allow for the surgical checks you can have in medical sciences, or applications as free as in the arts. If there is a field that evolves slowly and barely in pure theorisation, it surely is the HSS. To make it clear and simple, psychologists are still to this day Freudian, Lacanian or Jungian, sociologists are still pro-Bourdieu or anti-Bourdieu, linguists are still pro-Chomsky or anti-Chomsky, resulting in schools of thoughts in the last several decades that oppose each other in parish rivalries, which does not advance research in the slightest as those rivalries confine themselves to preserving the theoretical memory of an old figurehead. This attitude leads to thinking that HSS theories are neither flexible nor hybrid. They do have their own plasticity but they are not all flexible and hybrid.

If interdisciplinary work can make its theories hybridisable, which I strived to do in the second part of this chapter, a second problem that any interdisciplinary work faces is one we have previously discussed, definitions, and how to find an agreement amongst disciplines.

This was a focal point of Chap. 3 and I have already shown that it is possible to propose solutions to this problem, therefore we will not return to this. I would like to underline, however, the existence of a small variation of this problem that we have not mentioned yet – the problem of varying definitions, not from one discipline to another, but within the same discipline, in different languages.

As I come from the French research world, but mostly have had access to literature in English, it took us some time to realise that while "semiotics" and "semiology" were the exact respective translations of "sémiotique" and "sémiologie", the French terms did not cover the same exact fields as their English counterpart, and that, as the line was not at the same place in both languages, it was sometimes more prudent to understand and translate "sémiologie" into "semiotics", or vice-versa.

We can see that this second problem is actually two problems entangled together (both of which we have previously discussed): the problem of definition, of the proper meaning, and the problem of cultures (language-wise, historical, academic) that sustain different interpretations. It is therefore possible to list the three major problems that interdisciplinarity faces:

- The challenge of building a shared methodology (the hybridisation problem)
- The challenge of building with shared terms (the definition harmonisation problem)
- The challenge of building with shared referents and cultural baggage (the cultural gap problem)

Intertheoricity aims to allow building a model that takes into account all of these challenges and addresses them all with an identical method each time.

5.1.3 Guillaume's Theory

I will now show how Guillaume's theory addresses these various aspects of the issue of interdisciplinarity. When building the model for my semiotic methodology for animal studies, I will not strictly follow the whole of her theory, but I believe it is important to introduce it in order to clarify what we directly borrowed from Guillaume, and what I adapted to the special case that is our endeavour.

Firstly, Guillaume believes that not all the theoretical models necessarily need to be always evolving to be valid. Some barely are, others evolve more easily, and that does not presuppose their quality in the slightest. The problem is the need to recognise the theoretical models able to work in intertheoricity when necessary. This is what she calls elasticity, the capacity to form contact points with other models and have one or more points of the theoretical model migrate without denaturalising it:

> *Les théories sont des processus de pensée abstraits, souvent schématiques et schématisés, elles s'étudient en synchronie comme en diachronie. Certaines théories sont cumulatives (théorie des nombres), certaines sont évolutives, transférables et adaptables (F.Rastier), d'autres évoluent du fait de leur créateur, mais pas par osmose extérieure (A.J.Greimas). Les trois formes de théories sont complémentaires, et cette diversité est une richesse pour la recherche. Quand les théories sont monodisciplinaires, elles évoluent peu, quand elles évoluent, c'est qu'elles peuvent s'ouvrir ou s'appliquer à d'autres sphères qui vont s'emparer d'elles pour les transformer, elles deviennent alors de nouvelles théories, portant en elles quelques traces, ici et là, d'autres théories très anciennes ou encore très actuelles. [...] La transférogenèse, théorisation du processus de transfert de l'objet culturel, est un exemple d'interthéoricité et d'élasticité des théories. L'élasticité est un terme souvent employé en économie pour expliquer les variations d'une grandeur. La notion de variation par rapport à une théorie source est le point de contact qui permettra ici de faire migrer une partie de la théorie d'où son élasticité sur un (ou plusieurs) point(s) de contact, sa nouvelle plasticité et son hybridité finale.* (Guillaume 2014 p. 17)

> Translation:
>
> Theories are abstract thought processes, often schematic and schematised, and are studied in synchrony just as in diachrony. Some theories are cumulative (number theory), others evolve as their creator wants them to, but not through outside osmosis (A.J. Greimas). The three forms of theories are complementary, and this diversity is a boon for research. When theories are monodisciplinary, they rarely evolve, and when they evolve, it means they can open themselves to, or be applied to other spheres that will take them and transform them, and they will then become new theories, keeping here and there traces of other theories, some very ancient and others very current. [...] Transferogenesis, the theorisation of the cultural object transfer process, is an example of intertheoricity and elasticity of theories. Elasticity is a term often used in economics to explain the variations of a measure. The notion of variation in relation to a source theory is the contact point that will allow the migration of part of a theory, hence its elasticity on a (or several) contact point(s), its new plasticity and its final hybridity (Guillaume 2014, p. 17)

These three aspects that Guillaume mentions do not have to be present in equal parts – to have one of the aspects be predominant, or be that way at one point and then not as much or at all at another is normal and expected (see Fig. 5.1). However, I will not go into more detail on this aspect, as the part that truly interests us is the one on the migration of part of a theory to new contact points.

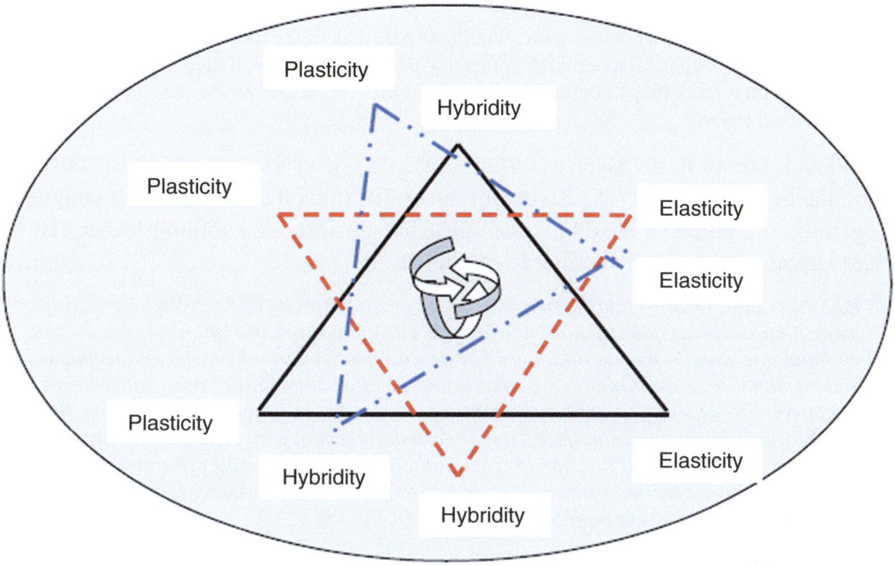

Fig. 5.1 Tri-triangular dynamic of intertheoricity, from the habilitation thesis of Guillaume

To know whether such a migration is possible, it is necessary to look once more at the challenges listed in the previous part and understand how it is possible to find a solution to these problems. As for the building of a shared methodology, Guillaume insists on the major aspect of modelling:

> *La plasticité des théories est en lien direct avec leur modélisation ou leur schématisation. Les théories sont toutes plastiques quand elles sont modélisées. Cette plasticité qui leur est propre, permet de reconnaître immédiatement la théorie de X ou de Y.* (Guillaume 2014 p. 24)

> Translation:
> The plasticity of theories is directly linked to their modelling or their schematisation. Theories all have plasticity when modelled. This plasticity, which is their own, allows one to immediately recognize the theory of X or Y.

In this sense, the Peircean triads are very suitable candidates for a shared methodology, as they are extremely modelled. Furthermore, they will work particularly well alongside theories that include very clearly schematised tools, like ethograms can be.

Concerning the definition problem, it is important to recognise that a terminology has a history (and I have deeply discussed this point in Chap. 3) and that history must be taken into consideration, whether we are trying to harmonise definitions or create a new terminology:

> *La terminologie qui accompagne les modélisations est la marque de l'époque qui les a vues naître, la touche lexicale et esthétique propre à leur siècle, la touche personnelle de leur auteur. Par le biais de l'interthéoricité, les modélisations et les terminologies peuvent se mêler et créer une nouvelle plasticité accompagnée de termes hybrides.* (Guillaume 2014 p. 24)

Translation:

The terminology that comes with modellings is a mark of the time they were created in, the lexical and aesthetical touch of their century, the personal touch of their author. Through intertheoricity, modellings and terminologies can mix and create a new plasticity with its own hybrid terms.

When it comes to the issue of culture, intertheoricity aims to take into consideration the fact that a theory is also the product of its time, its culture and its language. Logically, it cannot be absolute, and must not be seen as a monolith that can be called upon without ever needing to modify it.

> *Pour aller dans ce sens, nous ajouterons aussi que la démarche de sacraliser les théories ou de n'en suivre qu'une seule, est une attitude elle aussi restrictive qui n'est plus du tout en harmonie avec le monde d'aujourd'hui devenu multiculturel, hybride, multiethnique, multimédial, en transformation permanente, où l'on attend de tous un potentiel d'adaptabilité, de souplesse comme jamais auparavant dans l'histoire de l'humanité. Bien des théories restent, elles, statiques, comme d'un autre temps, dont elles proviennent indéniablement, surtout quand l'Antiquité et ses philosophes sont convoqués pour analyser une situation des plus contemporaines. Si la démarche peut sembler érudite, la transférabilité du propos est bien souvent risquée.* (Guillaume 2014 p. 28)

Translation:

To move in this direction, we will also add that the endeavour of sanctifying theories or only following one is an attitude that is also restrictive and completely out of touch with the world of today, which has become multicultural, hybrid, multi-ethnic, multimedial and constantly transforming, where we expect some potential of adaptability from everyone, a level of flexibility never seen before in the history of humanity. On the other hand, many theories remain static, as if from another time, which they are, undeniably, especially when people call back to Antiquity and its philosophers to study a very contemporary situation. While it may seem like a scholarly endeavour, the transferability of those remarks is often hazardous.

This is one of the aspects that I will refer to the least in the following part, for several reasons. The first is that the cultural issue is, in the case of a semiotic model focusing on animal studies, already very integrated in the terminology issue, which I have discussed in detail in Chap. 3. The second is that my difficulties actually come from what I could call academic cultures, that is to say, how to have tools and concepts from different disciplinary circles interact and communicate permanently and rigorously.

5.2 How Intertheoricity Allows for Bigger Models

In this part of the chapter, I will show how we built of semiotic model of animal studies with the help of the concept of intertheoricity. Each step will give me the opportunity to show in what way this concept solves one of the challenges mentioned in the previous part, and how, in the present case, we could build a "bigger" model than we otherwise could have.

Firstly, I will return to the crucial point of having a shared methodology when working with different academic fields. The example given here will focus on the coexistence between language sciences and life sciences on a broader level, and the coexistence of semiotics and ethology in particular.

Secondly, I will return to the previously discussed (in Chap. 3) problem of definitions. If I spent so much time on definitions before this chapter, it is because they proved to be a major issue, much too vast to be handled in a simple manner in the following part. I will confine myself to a reminder of the way the definitions I discussed in Chap. 3 are the result of an intertheoretical work that aims to reduce the misunderstandings between the involved disciplines. To do so, the model offers two radically different types of solutions: the harmonisation of existing definitions, and the creation of new definitions when the existing ones are incompatible.

Lastly, I will show in greater detail how intertheoricity leads to concepts of different theories "talking" to each other by uniting them in a single, harmonised and unified entity. By "talking", we mean exchanging elements, tools and functioning points from one field to another, without the barrier of definitions, methodology or academic traditions. Consequently, I will discuss how intertheoricity leads to the building of richer, bigger models.

5.2.1 A Shared Methodology

To build a semiotic methodological model adapted to animal studies, it is important to reflect upon the way to make a shared methodology out of several different ones. In my case, this focuses more specifically on the study of emotions, but the methodological model itself can be applied to any form of study of more or less complex semiotic acts in non-human animal species.

To do so, I needed to understand how to hybridise these methodologies, which, to begin with, means understanding what each of them needed to properly function. In order to dissect those needs, I created a three-level model that goes from with the most general to the most detailed, summed up in Table 5.1.

The first level aims to take in consideration the biological and semiotic variations in the subject species. This level is on a vast scale and broadly applies relevance tools, with the main goal of determining the frame of the study.

The goal of the second level is to locate the place that the semiotic acts we wish to study take in the subject species or group. This level aims to integrate the semiotic sphere within what already potentially existed in ethological study and to see how the two can interlock, regarding which aspects and what their limits are.

The third level is a more precise study of the targeted, precise or complex semiotic manifestations, linked to one or a few subjects and offers interpretations that will give a different perspective on the ethological view of certain aspects that are difficult to reach (cognition, emotion, memory, etc.)

Table 5.1 General summary of the model

Level	Goal	Steps
First	Take into consideration the biological variations in species	Adjusting the expectations of the experimenter to the biological limits and particularities of the studied species, identifying the experimenter's own cognitive biases, knowing how to adapt to the circumstances, living and testing conditions and the existence of the autonomous mental world of the studied individual.
Second	Understand the place of semiotic acts in the lives of individuals	Understanding the place of semiotic acts in the social, emotional and mental lives of the individuals, correctly interpreting the complexity, precision and motivation of their semiosis, offering ways to control and elements that could contradict the obtained conclusion or improve the given reasoning.
Third	Study complex semiotic manifestations	Studying the specific semiotic manifestations, knowing which category of conscience and will they belong to, placing them back into the environmental, historical and individual context of each studied subject, understanding the place that these manifestations occupy in their life and mental world and their degree of importance to the subject.

All three of these levels will be explained in detail in Sect. 5.2.3, when I will be discussing the interactions that the different tools and concepts have with each other.

In the first level, we find in the steps the particular attention given to short-circuit anthropocentric and anthropomorphic biases that I mentioned earlier. In this level, there are two centres of coexistence: one will form the traditional methodology of life sciences with relevance tools, and one that is created by those same relevance tools with the available ethograms. One centre aims to filter theoretical tools, the other to offer a model that fits the circumstances that the researcher has to deal with.

In the second level, the functioning steps will focus more on the place that the semiosis that will be studied holds in the life (in all its aspects) of the individuals. The centres of coexistence come from the various types of academic knowledge from life sciences (mainly ethology and neurology) with, on one hand, the tools to analyse semiotic manifestations (Peircean classification, first and foremost) and, on the other hand, the tools to analyse general semiosis (which mainly come from pragmatism). In the first case, the aim is purely descriptive in regard to the semiotic manifestations, in the second, the aim is interpretative, with the connection of signs, ideas and data.

Lastly, in the third level of this model, the functioning steps are finer and focused on the sign spec. On one hand, the point is to categorise the signs using the Voluntary-Conscious-Unconscious to offer a finer analysis of the processes occurring in the individual. On the other hand, these signs will also be weighed using the IFC grid, which will help placing the semiosis in the context of the individual emitting it (and potentially, of those that are meant to receive it).

Methodological elements from the different fields are in every step and play their roles in their field of expertise. As the type of model presented in this book is a semiotic study on animals, the semiotic tools are a creative force, while the tools from the life sciences act more as a safeguard and a methodological backdrop. There could easily be studies and models where the roles are reversed.

5.2.2 *Definitions: Harmonisation and Creation*

The second step to create this intertheoricity model, after the hybridisation of methodology, is to ensure that everyone means the same thing when a specific term is employed. To do so, as mentioned earlier, there are two possibilities: harmonisation and creation.

To avoid confusion as much as possible (as well as the translation issues we also mentioned earlier), I think it is always preferable to harmonise the definitions of employed terms rather than creating new ones.

As a result, in my thesis, there was a long thought process to put together a harmonised definition for "Emotion", the one I mentioned in the first section of Chap. 3, and one for "Memory". I see these definitions as being specific to this particular type of work on complex animal emotions, and not necessarily fundamental to all semiotic work on animals.

However, the definition for "Conscience" that was also presented in the first section of Chap. 3, is one I deem crucial to any type of work, especially since part of the model uses the conceptual grid of Intentional-Conscious-Unconscious, which is directly related to this notion. There is also another definition that I wanted to rework; the definition of the notion of "Ecology". Returning to this notion seemed important as its main, most well-known meaning is very different from its ethological meaning, which is the idea of an experiment, for example, respecting the biology of the individual going through a puzzle test. Boesch (2007) considers that test that forces young monkeys to solve puzzles while separated from their mothers is not ecological, as it does not take into account the stress factor which is a biologically normal thing for a young primate to feel when left alone.

When harmonisation is hard to reach, one should ask whether the concept is really necessary for the study, or if it can be substituted with a less controversial, more neutral concept. This is what I did here with, amongst others, the terms "Intelligence" and "Culture", discussed in the second section of Chap. 3. While interesting, especially the concept of intelligence, finding a viable definition was much too complex, and other terms could easily be used as substitutes without losing the original meanings while being more neutral (for example, "Cognition" replacing "Intelligence").

It may be necessary to have to create a new term as a last resort. When the concept itself is missing (as it was with the case of the intertheoricity concept), the creation of a new term fills a lexicological void. When the term is so controversial that it taints the rest of the work (as was the case of "language", which I replaced with "saying" as developed in the second section of Chap. 3). For more details, we direct back you to Chap. 3.

5.2.3 *How Concepts "Communicate" with Each Other*

If we wish to look more closely at the interactions of the different concepts from different fields, it is of interest to explain in detail the levels presented in Table 5.1.

Table 5.2 goes into the details of the first level of the model. It shows how relevance is a tool that adapts to different theoretical concepts (i.e. anthropocentrism), practical tools (i.e. the available experiment material) and experimental data (i.e. known data on the living conditions of the individual). It is the layering of the reading grids on these various methodological aspects of life sciences that will lead to only keeping the relevant tools for a precise study, and ensure a suitable model.

Table 5.2 First level reading grid

Level/Aim	Tools	Goal	Steps
First Take into consideration the biological variations in species	Relevance applied to anthropomorphism	Filter theoretical tools	Eliminating anthropomorphic expectations (e.g.: do not look for tears as a sign of sadness)
	Relevance applied to anthropocentrism		Eliminating anthropocentric reasonings (e.g.: this animal does not cry about a death in the group, therefore they cannot feel sadness)
	Relevance applied to tools		Choosing the tools most appropriate on a biological level (e.g.: studying long-term sadness using photographs, skulls, items related to the dead individual)
	Ecology of social ties	Provide a model that adapts to the circumstances	Taking into consideration the biological social functioning of the animal (e.g.: they only show sadness to individuals they know, are related to them and/or are in their group)
	Ecology of living conditions		Taking into account the living conditions and their influence on the individual (e.g.: this animal, raised in captivity, manifests sadness in empathy with other animals in their group, but also their carers)
	Ecology related to the notion of individual		Taking into account the individual and their past experiences when looking at their emotional reactions (e.g.: this animal is not sad at all about the death of a member of their group with whom they had a violent and open conflict)

Table 5.3 presents the second level of the model. Its two main lines (descriptive and interpretative) are distinctly separated and clearly show the two aspects of this level. On one side, the Peircean classification, the notion of sign precision that stems from it and the concept of combinatorics (which can include supersigns) aim to describe the manifestations. On the other, pragmatism establishes a connection between the sign and the idea that will allow for the interpretation of the semiosis. Note that the last level of this grid is a safeguard; its role here is to provide other possible interpretations and put forward semiotic testing paths to decide between interpretative hypotheses.

Table 5.4 presents the last, finer level of the model. Its purpose is to more precisely study the signs described earlier, which have already had their global semiosis

Table 5.3 Second level reading grid

Level/Aim	Tools	Goal	Steps
Second Understand the place of semiotic acts in the lives of individuals	Peircean classification	Describe the semiotic manifestations	Understanding the level of complexity of the emitted sign (e.g.: Koko emits plaintive sounds when All Ball dies, it is a simple sign; she also signs "cry", which is a symbol, and therefore a complex sign)
	Notion of sign precision		Understanding the degree of precision of the emitted sign (e.g.: Koko has a specific sign for All Ball, her expression is directed towards a very precise subject
	Sign combinatorics		Understanding the meaning of the combination of emitted signs (e.g.: Koko immediately violently reacts to the combination of "All Ball – Dead", this succession of signs shows her understanding of the gravity of the situation and her strong emotional state)
	Pragmatism of thought	Connect the sign, the idea, the emotion and the data	Interpreting the thought translated by the sign (e.g.: the immediate, repetitive reaction from Koko makes us think she understands the definitive nature of the separation and that it is extremely painful to her)
	Pragmatism of production		Understanding the motivation of the semiosis (e.g.: by expressing her sadness, Koko ensures that she will receive the support and comfort of Penny)
	Pragmatism of error		Studying the analysis of potential errors and how to test them (e.g.: if, in the following days, Koko had asked to see All Ball, it would have made it clear that she did not correctly understand the concept of death)

Table 5.4 Third level reading grid

Level/Aim	Tools	Goal	Steps
Third Study semiotic manifestations, including some complex ones	Voluntary signs	Categorise the signs	Recognising and interpreting voluntary signs (e.g.: the male chimpanzee helping the wounded female with carrying her baby)
	Involuntary signs		Recognising and interpreting involuntary signs (e.g.: he tries to hide manifestations of distress when the rest of the group moves a bit too far away)
	Unconscious signs		Recognising and interpreting unconscious signs (e.g.: he has very brief physical reactions every time the female lets out a plaintive cry)
	Nuance of context	Weighing the signs	Placing the semiosis in its context (e.g.: the male is not the partner of the female, or the progenitor of the young, his action cannot be explained as done out of immediate sexual or reproductive interest)
	Nuance of frequency		Properly interpreting the place that the semiosis has in the life of the individual (e.g.: the male makes sure to carry the infant of the wounded female several hours per day, for days, which shows his involvement)
	Nuance of intensity		Interpreting the semiosis in the light of its marks of intensity (e.g.: the action of the male is costly energy-wise, difficult, possibly dangerous, and does not give him much in return; there needs to be a strong internal move to make him act this way)

more or less interpreted. The aim here is to categorise signs according to the Intentional-Conscious-Unconscious models, in accordance with known observations and data from the relevant specialties in life sciences (primatology, neuroscience, cetology, etc.). These signs will then be nuanced through the use of both the IFC semiotic scale and the existing field data. The perpetual dialogue between these two sides leads to a relevance that is as close as possible to the obtained results.

5.3 Why We Need Bigger Models

To conclude this chapter, I want to look more closely at the necessity of intertheoricity; why do we actually need larger models to answers questions that, after all, already belong to disciplinary fields that have more than proven their worth?

I split the reasons into three categories. While I do not have the pretention to say those are exhaustive, I do think the vast majority of cases that require a model that is larger than the traditional, individual academic models belongs in one of those three categories.

Firstly, I will deal with what is the majority of cases in our type of research; the more complex subjects that use relatively precise or fine aspects from various fields and that cannot be handled properly without them. This is the standard example of a study where interdisciplinarity, through intertheoricity, helps bringing more precision and more detail to a subject that is hard to study in a monodisciplinary manner.

My work is also representative, although not as much as with the first category, of a second type of case that requires the creation of a vaster model for it to be studied. As researchers become more and more specialised, it is sometimes difficult to find a methodological model that works on relatively "unexplored" subjects that have barely been studied (this was the case during our thesis about the zoosemiotics of complex emotions in non-human animals).

Lastly, there is a third category I thought important to make explicit, even if it may seem more down-to-earth. The case of having one study subject for which results and progress are expected in several research fields without having the possibility to allocate enough human, financial or material resources, as well as energy and time, into separate projects in their own respective discipline. In such cases, intertheoretical models become a way of grouping resources together and produce a shorter, less bulky research with a bigger impact.

5.3.1 More Complex Subjects

For a subject such as the one of my thesis, it seems difficult to only work with just language sciences, or just life sciences, without invoking the other at some point.

Subjects like this are called "complex" in this chapter, not in the sense that they are more intellectually subtle or need more advanced experiments, but rather in the sense that the fields most appropriate to handle them are much more strongly intertwined together than in monodisciplinary subjects.

I return to one of the cases I discussed at length in my thesis, the semiotic analysis of the expression of sadness, and mourning mechanisms, ritualised or not. In my opinion, it is complex, if not reckless, to decide that one academic field would be the only one able (or mostly able) to deal with this type of mechanism. But if we were to do this, which field would it be? It could possibly be psychology, specifically clinical psychology, which has spent a lot of time working on that subject in the past few years, but is rarely called upon (mistakenly, in my opinion) when it comes to non-human animals. I would rather propose the use ethology to study these categories of beings, but the subject of emotion is a tough fit for the usual ethograms, and a tougher one when the goal is to study the evolution of an emotion over several years (or sometimes decades, in the case of species with a long lifespan and a particularly well-developed long-term memory). If the study is aimed more precisely at the emotion itself, neurosciences should not be dismissed as they can provide its visualisation. But endocrinology should not be forgotten, as it also is a major component of emotions and their study. Semiotics can surely provide the most complete external description, but to do so, it needs the data and observations from primatology, cetology, ornithology, etc.

I could keep adding more and more to that list. But it is already obvious that it cannot provide a satisfactory answer, therefore I will stop here. Many disciplines can be involved in a single topic and competently deal with their preferred aspect. In the case of a specialised work, this can be perfectly sufficient, and it was, for a long time.

But there are two cases in which these subjects that have strongly intertwined disciplines absolutely need an intertheoretical approach at the risk of being otherwise vastly misunderstood, and those cases are logically at the two ends of the same line: either when the goal is to try and obtain a general model, or when trying to take a closer look at extremely detailed points.

In the first of these two cases, which is the one I worked on, as well as the type of work I am presenting and for which I am providing a methodology in this book, it certainly appears to be extremely difficult to have a general model for an interdisciplinary issue that would only include the point of view, theories and methods of a single discipline. I think that many mistakes that were made in the past were due to this method bias, which comes directly from the current over-specialisation of researchers, something we will discuss next.

This is the more delicate case because it entails building a model that is perfectly representative when it comes to balancing the theories and amount of involvement from each discipline in the issue. I hope to have managed to do so in the building of my own model: if the majority of tools are from semiotics (I am studying emotional manifestations after all), all of the data and observations were collected, verified and interpreted at first by researchers from life sciences, who I deem much more competent in this regard; while my tools connect the observed signs to the possible ideas at the source of their production, it is the life sciences that establish the cognitive prerequisites for the existence of such a process and tell me whether it is relevant to look for them in the subject species. This work would not have been possible without intertheoricity.

In the second case, which is more of a case of practical application, the problem can be more subtle. A work on an extremely specific point is usually picked by a researcher that is competent in a certain field specifically because this detail falls under the umbrella of said field.

It would be misleading, however, to think that the influence of aspects from other academic fields is non-existent simply because we are dealing with something on a very small scale. While this influence or intertwining exists in large models, it still exists, although possibly barely at all or marginally. If this influence happens to be minor, not taking it into account may be regrettable on a scientific level but still only have a small impact. But in some cases, that influence turns out to actually be a major one and can have large consequences regarding the exactitude of the given conclusions. For example, I gave in a colloquium the special example of the impact of affection on reproductive programs: while avoiding consanguinity is important, separating individuals with affective ties and long-term memory can turn out to be more harmful for the birth rate than not doing at all.[1]

5.3.2 *Over-Specialised Researchers*

The second reason that can make working with intertheoricity necessary in certain studies and certain subjects has already been briefly brought up in the previous point. It is an aspect that my own research was also faced with, and we think it is interesting to return to it, as it is representative, in our opinion, of a structural problem in research nowadays. While I have neither the intention nor the pretention to change this structure, I do think acknowledging the existence of this issue is necessary.

The rather new inclination of favouring interdisciplinary works, while not always implemented properly, has opened the path to new research fields on hybrid subjects that were often dismissed due to a lack of easy categorisation in one disciplinary field or another.

These subjects are still in their "exploratory" phase as a result, and, despite the age of its founding texts, zoosemiotics is one of them. There are few (although I can point out some exceptions) extremely comprehensive studies in this field (Mäekivi 2018), few methodological models (Delahaye 2017), and little theorisation (Rattasepp 2018). I, of course, hope to contribute to this list through this very book, amongst other things.

But beyond that inclination, it is clear that this lack of documentation has several origins and the current over-specialisation of researchers is one of them. It leads to a certain hesitation towards those "exploratory" works that demand general studies, skills from multiple fields and a very demanding interdisciplinarity. The amount of work can easily be overwhelming, and many can refuse this type of subject as they feel lacking in skill and knowledge about these aspects.

Thankfully, nobody nowadays simply works in science on their own, isolated from their colleagues, and the problem can easily be circumvented by gathering an interdisciplinary team where each member steps in when their specialty is involved.

But in reality, these teams are fairly rare, and it is easy to understand why. With the over-specialisation in knowledge and know-how comes the over-specialisation in methodology and theory. Such teams often have a lot of trouble working together if their members come from wildly different academic – and consequently wildly different theoretical and methodological – circles.

Yet for the study of our type of work, it is necessary for linguists, and veterinarians, and lexicologists, and ethologists, and semioticians, and biologists to all contribute together in one model.

---

[1] P. Delahaye "To Care, to Look After, to Protect: Semiotics of Benevolent Violence", for Hybrid Natures, University of Tartu, November 2018.

This is where a good knowledge of intertheoricity will prove useful when it comes to circumventing the differences in methodology, theories and terminology. It will allow building a model fitting this type of hybrid work, which often requires methodological and lexicological creations. Without this type of framework, it is probable that research quickly becomes laborious, due to a lack of understanding, or rigor, or harmonisation. Without intertheoricity, hybrid subjects that are still being explored remain crippled by the over-specialisation of researchers and the lack of shared methodology.

5.3.3 More Impact, Less Time

There is one last reason that justifies the need to include intertheoricity in a study. I introduce it here in a more hypothetical way than the others because this aspect did not specifically concern my work.

There are times where, on a single interdisciplinary study subject, different results and progress are expected in the different fields involved. While this point may seem more down-to-earth than the other two, a healthy dose of pragmatism would lead one to admit that it is not always possible to allocate enough human, financial or material resources, as well as energy and time, into separate projects in their own respective discipline.

And as I mentioned earlier, it is not always advisable to separate the disciplinary fields that are so tightly intertwined, and can be harmful most of the time.

The intertheoricity model adapts particularly well to what is usually seen as a flaw in the current research system: the ever stricter time constraints. Some years ago, a thesis in human sciences would often take four, five, maybe even six or more years to be completed, the more and more drastic reregistration directives tends to force young researchers in all fields to finish their first research work in 3 years, maybe four at most.

But they are far from the only ones affected by this; even the most seasoned researchers have to live by the "publish or perish" policy to a certain extent, which forces them to publish often and regularly, leaving little time for actual research.

The intertheoricity model allows the study of several aspects of the same subject by adding elements that can be exploited by several disciplines which causes the research to have a bigger impact in a lesser amount of time spent (as well as a potential reduction in human and material resources needed, since the research consolidates needs that would have otherwise be split amongst several studies, although this remains, for now, hypothetical on our end).

It is very much possible – and that is the case of the author of these lines – to find the current state of research and well as this way of functioning to be deplorable. Nonetheless, it is a reality and intertheoricity can bring solutions to help deal with it.

Whether it comes down to a real difficulty or a policy that the current state of research dictates, intertheoricity allows the building of models with which studies can create a bigger impact at a lesser cost in research time, provided the intertheoretical methodology is solid and the resulting model is rigorous. It should be noted that on this point, semiotics is a very well adapted discipline: while it is now in human sciences, amongst language sciences, it long belonged to life sciences, mostly medical sciences, and still keeps their codes. When a team wishes to build an intertheoretical methodology, semiotics should always be suggested as a discipline that "interprets" the others.

References

Boesch, C. (2007). What makes us human (*Homo sapiens*)? The challenge of cognitive cross-species comparison. *Journal of Comparative Psychology, 121*, 227–240.

Delahaye, P. (2017). *Etude sémiotique des émotions complexes animales : des signes pour le dire*. Paris: Paris Sorbonne – Sorbonne Université.

Guillaume, A. (2013). *Transférabilité du sens d'hier et d'aujourd'hui. Des mots, des signes, des cultures*. Nanterre: Université de Paris Ouest La Défense.

Guillaume, A. (2014). L'interthéoricité : sémiotique de la transférogenèse. Plasticité, élasticité, hybridité des théories. *Revue PLASTIR, Plasticités, Sciences et Arts, 37*, 1–36.

Hess, J., Beever, J., Strobel, J., & Brightman, A. (2017). Empathic perspective-taking and ethical decision-making in engineering ethics education. In D. P. Michelfelder, B. Newberry, & Q. Zhu (Eds.), *Philosophy and engineering: Exploring boundaries, expanding connections* (pp. 163–179). Springer. https://doi.org/10.1007/978-3-319-45193-0.

Langer, M. C., et al. (2017). Untangling the dinosaur family tree. *Nature, 551*, E1–E3. https://doi.org/10.1038/nature24011.

Mäekivi, N. (2018). *The zoological garden as a hybrid environment – A (zoo)semiotic analysis*. Tartu: University of Tartu Press.

Michelfelder, D. P., & Newberry, B. (2017). *Philosophy and engineering: Exploring boundaries, expanding connections* (pp. 163–179). Springer. https://doi.org/10.1007/978-3-319-45193-0.

Rattasepp, S. (2018). *The human mirror. A critique of the philosophical discourse on animals from the position of multispecies semiotics*. Tartu: University of Tartu Press.

Chapter 6
Strengths and Flaws of Ethological and Biological Methodology

Abstract Creation sub-question n°2: what are the parts, strengths or flaws, of the pre-existing animal study methodology upon which our model must be built?

This chapter will present the points of hybridization of the semiotic model with life sciences, starting with the strengths. It will thus get back to the seniority and the methodological solidity of the field of life sciences, to what makes its methodology particularly evolutionary and to the fact that semiotics adapt especially well to a science based on observations. The chapter will then examine the weak points, including the controversy about the place of the laboratory in the studies on complex aspects of animals. This second part will also tackle the issue of disruptions related to observation, as well as the issue of awareness of ideological positions, still relatively strong among this disciplinary field.

Keywords Biological methodology · Ethological methodology · Intertheoricity · Interdisciplinary models · Ideological biases · Methodology of observations

In this chapter, I will discuss how my intertheoretical model fits into life sciences and their methodology. In the previous chapter, I showed how it was made to integrate elements from life sciences and work with its theories, but I did not show how the model itself fits into life sciences, especially on the subject of animals.

This chapter will be the opportunity to answer the second and last creation sub-question from my introduction: What are the parts, strengths or flaws, of the pre-existing animal study methodology upon which our model must be built?

This chapter, logically, is in two parts, each dealing with one aspect of the question: the strengths of the pre-existing methodology with which to work on one hand, and the weaknesses of the very same methodology that we need to make up for on the other.

In the first part, I will deal with the integration of our methodological model into the strong points of the existing methodology within life sciences on the topic of animal studies. I will discuss the fact that life sciences are an ancient, tried and true disciplinary field that has long thought about its methods and is particularly strong nowadays as a result. I will address the fact that, thanks to the omnipresence of the theory of evolution as a backdrop, life sciences are consequently an ever-evolving

P. Delahaye, *A Semiotic Methodology for Animal Studies*, Biosemiotics 19,
https://doi.org/10.1007/978-3-030-28813-6_6

discipline and the methodology can evolve with the various discoveries being made (which, for intertheoricity, makes life sciences an excellent candidate for theoretic hybridisations). Lastly, I will discuss the fact that life sciences are also sciences of the sign, sciences that are very much built on observations and with which semiotics can work particularly well.

In the second part of the chapter, I will look at the other side of that coin – the weaknesses of the existing methodology, and the ways my methodological model can provide tools to compensate for them. I will discuss the subject of whether researchers should leave the laboratory and the problems that come from both answers to this question. I will go more into detail about one of its underlying facets: observation is necessarily a disruptive element that is still not always treated as such. Lastly, I will return to this omnipresence of evolution theory, and how life sciences can be filled with a deep ideology, one that is not always easy to identify or circumvent, regarding the place of humans in the animal kingdom, and how this ideology can directly influence research and experiments.

6.1 Strengths to Work with

In this first part, I will take a closer look to the way a semiotic methodological model can be built in collaboration with life sciences. As my model focuses on animal semiosis, ethology will be the most solicited branch amongst life sciences in this part, but not the only one (in which case the other branches will be mentioned).

While longevity is not always a token of quality, we will begin with a study on how, in this specific case, the long existence of an academic field protects it from certain errors and makes the building of an interdisciplinary model easier and more fluid.

More precisely, I will show how this longevity is taken into consideration in my methodology and proves to be one of the scientific guarantees of the data with which my model will work.

I will then move on to the evolving aspect of life sciences, how this aspect impacts their methodologies and how they can be useful in our work.

This point will have its counterpoint in the next part of the chapter; although the theory of evolution is omnipresent in life sciences nowadays, this was not always without friction, and some ideological reluctances continue to coexist with the field's methodological ability to evolve.

Lastly, I will talk about observations and how these sciences build themselves around them first and foremost. Observations are essential in a semiotic model as there can be no description and certain no interpretation of the sign without a situation in which signs are emitted.

I will show how methodology building around observation from life sciences is an opportunity for semioticians that wish to study animals and I will give some examples (which will logically be more focused on the emotional expression aspect) from my work to feature how the methodological model thrives on observations from ethologists.

6.1.1 Ancient and Strong Field

Historically speaking, the limits of life sciences were not always as clearly defined as they are now. Philosophers would often try their hand at classifying species and naturalists were sometimes men of the cloth. While it would be impossible to trace the entire history of the field in so little time here, I will simply take a closer look at the evolution of the works and theories that are directly related to our subject, and at how they changed their methodology over time. Most of those works have already been quoted in previous parts to present or clarify this or that point.

The first studies on animals were clearly not what one would call "scientific works" according to the standards of today. Aristotle's classification of species (Bertier 1994) was more of a philosophical work than a scientific one, as were the works of Augustine of Hippo later on (de Fontenay 1998, pp. 365–374). This can be explained by the fact that animals were not being studied to be understood, but to find out what made humans so special. This bias lived on in human sciences (Rattasepp 2018) but life sciences slowly corrected the course by beginning with studying animals more and more simply for what they were, starting with Buffon (de Fontenay 1998, pp. 567–568) and reaching maturity (but not its end) with Darwin. With him (and other researchers that are more or less his contemporaries, but it would be far too complicated to draw up an exhaustive list), modern ethology and evolutionary biology truly begin.

While we can attribute the true ethological methodology (vitalism and behaviourism excluded) to Lorenz (1967), the interest given to animal behaviour and, more specific to my case, animal emotions comes from the very origins of modern ethology and evolutionary biology. I did not deem possible to put together a review of theoretical ethology without including the works of Jakob von Uexküll (1934), for ethology, and Charles Darwin (1872), for evolutionary biology (as he is its founder), particularly *The Expression of the Emotions in Man and Animals*, undoubtedly the first real study in the field.

I am not simply mentioning a pioneer to point out his work as a founder, but also to remind some stances that were his own and were later forgotten or pushed aside, notably by behaviourists, and that have started to resurface recently at the instigation of some ethologists, primarily primatologists. These stances make the theories of life sciences even more easily hybridisable, and therefore easier to work with others in an interdisciplinary work.

Marc Bekoff (2009) took on the role of defender of the Darwinian stance on two different, but connected, points. The first one, which I already mentioned, is that anthropomorphism can be seen more as a fine knowledge-acquiring tool rather than an experimental or theoretical bias (Bekoff 2009, p. 236). It is the result of both our cognitive evolution and the emotional capabilities we share with certain branches of the animal kingdom. This point is of particular use to us in the second level of our model (Table 5.3), in which the interpretation of semiosis raises the question of anthropomorphism.

The second point comes directly from Charles Darwin's experimental methods and processes, as we will explain in detail in the next part. According to Bekoff and other scientists from the same school of thought, it is the importance of preferring observation (and therefore the accompanying field reports) to experimentation in a controlled environment (*Ibid.* p. 82). The idea here is that observation will always lead to more ecological data than experimentation, which is in a controlled environment that logically is very artificial to the test subject.

However, I did not think it wise to exclude any form of experimentation. Instead I chose to have a hyper-critical eye on all of parameters that could potentially influence results (the entire next part about the connection of results to the laboratory and the observer comes from this choice). But my firm preference for observations made me opt for a multimodal corpus that includes very diverse multimedia content (see Sect. 1.3.2) despite all the various difficulties the analysis of this type of contents entails in the context of a scientific study.

The conflict between different schools of thought in ethology has led to very strong, opposite stances (the exchanges between De Waal and Ghiselin (De Waal 2013, p. 49) come to mind). When it comes to the choice of methodologies and theories within ethology and its own schools of thought, I mostly rely on two authors (particularly two of their books) to know how to work, with which schools and which theories from life sciences to draw from.

The first author allows an interdisciplinary work that is very much in tune with the current data, methodology and schools of thought in ethology. I am talking here about Peter Machamer (2009), particularly his works for *The Oxford Handbook of Philosophy and Neuroscience* (Bickle 2009), already mentioned earlier. In those works, the author positions himself firmly against behaviourism, not just because he exposes its methodological flaws and deficiencies, but mostly to warn against the temptation of content oneself with just using its models and explanations, which are insufficient to deal with the complex issues that stem from particularly evolved species. He also warns against the beliefs that there is no need for anything else in the study of animal behaviour. Which is a promising starting point when offering other models, methodologies and tools.

The second author is a special case that I will study in more detail in the next point, when talking about the evolving methodology of life sciences. Because if this field is so strong nowadays, it is specifically due to the fact that it evolved, sometimes very radically, throughout its history, and does not close itself off to new possible theories and methodologies.

6.1.2 Evolutive Methodology

The second author I wanted to mention allows us to take the stance of "coexistence" with language sciences and lays the foundations of the functioning of an interdisciplinary work such as mine. He is unique in the sense that it does not come from life sciences, strictly speaking. I am talking here about Thomas Sebeok (1976), specifically this time about *Studies in Semiotics: Contributions to the Doctrines of Signs*, and the explanations the book brings on the complex intertwining of biology, language and culture (Sebeok 1976, p. 86). By explaining how these different fields fit into each other, Sebeok opens up new possible research paths on the way the disciplines studying them can work together.

Different studies also highlight this evolving aspect of the life sciences methodology: many observations and field reports can be regularly re-interpreted through the prism of other discoveries, sometimes from other academic domains. As it is not possible to offer an exhaustive list here, I will instead present the works I used during my research to show how each of them influenced the methodological model.

It is important to point out that these are the "major" studies in our corpus, and some points will require additional works to refine our remarks on a certain aspect, or bring up a counter-example. These works will be brought up from time to time but not be the subjects of an extended, deeper study.

One of the flagship studies that I often used as a reference point is from Bodamer and the Gardners (2002), *How Cross-Fostered Chimpanzees Pan Troglodytes) Initiate and Maintain Conversations*. When faced with the considerable difficulty of trying to understand fundamentally foreign minds, it seemed necessary for us to lean on a study focusing on our close cousins (and therefore most likely focusing on the least foreign minds we could have to face).The methodologies from the Gardners as well as the works from their assistant, Fouts (Fouts and Mills 1998) will also be used as reference tools to read and understand the works of other ethologists, especially primatologists.

I included in my corpus, for the same reason, the book by Marc Bekoff (2009). While it was not a study report in and of itself, it did include several that often contained notes, comments, and were connected to other studies in the same field, giving us a slice of expertise in ethology, which, as a non-specialist, we could have lacked. His work is a very good example of how certain observations, including old ones, can be given an entirely new interpretation years later thanks to new knowledge, as long as they were originally described with enough precision (we will get back to the importance of observation when collaborating with life sciences in the next point).

In the same viewpoint but from a different disciplinary field, I have included the book of Borsberg, Burley and Hämäläinen (2012), *Language, Ethics and Animal Life*. Along with the compilation of various, abundant studies and the diverse perspectives on these works, this book is of interest to us in this sense because it presents a certain amount of research from ethologists and looks at them using the perspectives from the theories and works of linguists (this is the case, for example, of Terrace's experimentations seen through the prism of Wittgenstein's theoretical writings, a case we mentioned previously in Sect. 1.3.1).

Afterwards, other studies were taken into account in my corpus and studied using my model. They were looked at through the prism of previously mentioned books, as well as the *Manuel d'Ecologie Comportementale* (Danchin et al. 2005).

There are several reasons for this. The first is the use of this handbook lets us take into account aspects of ethological studies that rely on very advanced notions (from biology, neurosciences, ethology, etc.) such as the notion of culture:

> [*citant* Freeberg (2012) *pour étude sur la culture] « l'ensemble des traditions comportementales d'une population qui sont apprises et transmises socialement » Le terme de « tradition » insiste sur la nature nécessairement conservatrice de la culture, au moins sur des bases de temps courtes ou des échelles spatiales réduites.* (Danchin et al. 2005 p. 46)

> Translation:
> [quoting Freeberg 2012 about the study of culture] "the whole of behavioural traditions of a population that are learned and transmitted socially". The word "tradition" underlines the necessarily conservative nature of culture, at least on a short-term basis or on smaller spatial scales.

The second reason for the specific use I have for this book comes from the considerations its authors have about theoretical models, their conception and the way they work:

> *Par définition, un modèle représente une approximation du réel. Il existe cependant divers degrés de sophistication d'un modèle [...]. D'une manière générale, un fort réalisme pour un modèle en limite la généralité et en diminue la simplicité et donc la valeur didactique.* (Danchin et al. 2005 p. 52)

> Translation:
> By definition, a model represents an approximation of reality. However, there are several degrees of sophistication of a model [...]. Generally, a thorough realism in a model limits its general aspect while diminishing its simplicity, and as a result, its didactic value.

This would indeed help us push our model from the start in a certain direction, depending on the purpose we want to give it. In the current case, the aim was for me to provide a model that could easily be generalised, as well as have an important didactic impact. As a result, I knew from the start that I would need to make it as simple as possible during my work, and to try not to make the parameters too specific and realistic, unless I were to have different levels that each had different degrees of simplicity and precision.

This type of functioning explains the current form of my methodological model: level 1 (Table 5.2) can be easily generalised but is almost not detailed at all, while level 3 (Table 5.4) can barely be generalised because it includes realism parameters that are much more specific. The form of my methodological semiotic model, in different levels, is a direct consequence of the methodological functioning of life sciences and the hybridisation between them and semiotic methodology.

6.1.3 Observation-Based Science

Being based on observation makes biology, on a general level, and ethology more specifically, particularly interesting from a semiotic perspective, as it shows it is a discipline that is also focusing on the description and interpretation of signs. Even better, many of their studies that are very well documented end up describing phenomenon they were not originally looking for, with a fortuitous observation, or one with an impact that was not immediately comprehended by the research team (sometimes just because it was not their study subject).

Therefore, when working on animals from a semiotic perspective, the first phase can be achieved through a corpus of observations that had already been reported by other teams, sometimes with a wide variety regarding species, eras or goals, that coincidentally end up describing similar elements.

During my thesis work, when I had to test my methodological model, it would have been impossible to do so on several species, in several given cases, just to observe the relevance of the semiotic reading grids on an emotion. But by compiling the observation data that had already been described, it was achievable.

To illustrate this, I will give the example of the first part of my study on emotional trauma, which led me to think that this trauma was the state of a subject whose mental world was overwhelmed by its own manifestations – that is to say, by themselves. To support this point, I gathered observations from cetologists, primatologists and researchers that specialised in the study of elephants.

Whether I look at the reports from Fouts (Fouts and Mills 1998), Patterson,[1] Moss (1989, pp. 75–77) or Gonzalvo, trauma always seems to catch the individual off-guard and reach an unprepared mental world. Gonzalvo has some particularly striking words when he explains that particularly difficult mourning processes, the ones that can even become pathological, can rarely be found when the individual had been expected to die for some time. However, in the case of a brutal, unexpected death, these processes are blatant and explosive.

When placing my reading tools on the semiotic manifestations of an individual in whom ethologists suspect a trauma, striking similarities appear in terms of intensity, frequency, and types of emitted signs:

- Behaviours revealing the impact of trauma on the subject have an incredibly high frequency, to the point where they are almost permanent in the sense that there always seems to be at least one sign of the severe emotional state of the subject (prostration, complaints, avoidance of others, a frantic search for contact with other members of the group, etc.)
- The intensity of the emotion(s) felt is such that it interferes with the life of the subject, even in their most basic actions (feeding or cleaning themselves, etc.). Secondary actions (hunting, playing, etc.) often are completely put aside, at least temporarily (Bradshaw 2004)
- The subject has less control over their semiosis than they normally do, the emission of involuntary signs increases greatly, mostly signs of distress or dejection.
- Parallel to that is the subject seemingly having less control over voluntary signs, and indulges into bursts of anger, requests of affection to other members of the group or on the contrary rejections of others, much more frequently than usual.

[1] In this case, it was a scientific recording for teaching purposes. Part of this recording can be found on the website of the *Gorilla Foundation*, under the title of "All Ball".

- Lastly, when the traumatic event affects an individual that is part of a group, we can observe some changes in the semiosis of other individuals towards the affected one (more attention, requests, tolerance towards certain behaviours, or as Gonzalvo or Moss report, signs of assistance and observation of the perturbed subject).

What can I conclude, what interpretation can I have of signs that are so flagrant, so similar, in species, individuals and situations that are so diverse?

First, it is noticeable that the individuals producing this type of semiosis are gravely dysfunctional: they are incapable of feeding, protecting or taking care of themselves, they behave in ways that are dangerous while in the wild (long prostrations, loud complaints, erratic movement, etc.), there is no doubt that if this state were to continue existing, and if they were not in a group that, as mentioned above, tends to watch them closely during this phase, these individuals would be incapable of surviving in the wild, let alone reproduce and raise children.

This type of semiotic manifestations shows a lack of ability for the mental world of the individual to understand, integrate or deal with the event that occurred, and that is even truer for the impact the event had on them, the emotions they produce, and the way they feel. It is such a powerful emotion, accompanied with such a strong feeling, that the individual seems completely unable to handle while also remaining in a fully functional state. Many of their basic behaviours are gravely altered as a result, and they show signs of great confusion.

As I explained when discussing sign combinatorics, just because the semiosis is muddled or contradictory (such as when individuals are aggressive towards others then ask them to comfort them) does not mean it does not say coherent things. Here, the semiosis is a sign of the incapability of the subject to silence their emotions and of the dominating place it holds in their mental world. These signs tell us about the existence of a powerful and complex emotional life that can even, when brutalised, produce feelings that compromise the ability of the individual to live normally.

From a semiotic point of view, we can consider that there is a trauma in an individual when they show all the signs of a mental world that is overrun by its own emotional experiences, to the point of short-circuiting all the other aspects of the life of the individual. The trauma is also proof of the absolute priority that emotions can take for an individual when they reach a certain degree of intensity, including, in the worst cases, reaching a higher priority than their own survival instinct.

But this conclusion is only possible because the semiotic pattern can be observed in several species with ethological similarities, in several individuals from these species and in several standard situations showing similarities. The semiotic model is functional, but it is the massive amount of pre-existing data within ethology that gives it its necessary corpus.

6.2 Flaws to Counter

In this second part, we will focus on the faults and challenges that methodologies from life sciences can contain. This part of the chapter has two purposes; first, to acknowledge these faults so we can avoid them during an interdisciplinary work, and second, to offer – in our case, semiotically oriented – solutions to these challenges.

I will first discuss a dilemma that presents itself to researchers, especially ethologists: whether they should leave the laboratory. The experiment that can be reproduced in a laboratory environment has long been the archetype of the canonical experiment in life sciences, but its relevance is put into question in the cases of behaviour observation, particularly when the behaviour is complex or must be observed for a long period of time.

My second point will be the logical next step of handling the issue of observation; because observation is disruption. As a result, at a certain point, there is no observation that is perfectly objective or neutral, especially with complex animals: the researcher has to have a certain level of involvement with his study object, and we will look at the propositions that can be made so that this mechanism impacts the experiment results as little as possible.

Lastly, in order to present the counter-point mentioned in the first part of the chapter, I will take a closer look at the issue of the ideologies of life sciences. While they are on the front lines when it comes to set humans aside from other species, explaining the origin of our species or some of our capabilities, life sciences have a long and heavy history of ideology that can be hard to get rid of. I will show what this entails, and how to diminish the effects.

6.2.1 *Leaving or Not Leaving the Laboratory*

While it seems fairly obvious to study in laboratories when it comes to cell growth, resistance to antibiotics or the changes that come with this or that modification of the genome, other aspects of life sciences raise some methodological questions. What type of tests, which experimentations, what protocols should be used to understand language, emotions, thought? I had the opportunity to tackle this question in detail (Delahaye 2017).

To answer this question, it is of interest to look at the two opposite, previously mentioned cases of project Washoe (Bodamer and Gardner 2002) and project Nim (Terrace 1987). Both cases are projects in linguistics and comparative psychology on chimpanzees using sign language. The goal in both cases was to teach the simian subject signs from that language, and control their learning. Project Nim was a total failure (Forsberg et al. 2012) while project Washoe exceeded all expectation (Morin 1974). What happened?

Let us look at project Washoe first, since it was the first of the two to see the light of day. Led by the Gardners, Allen and Beatrix, and later on taken over by their assistant, Roger Fouts, the project was about teaching ASL to young chimpanzees, the most famous of which would be Washoe. Raised in a caravan, like a child, with toys, and pictures books, social interactions (see Fig. 6.1), walks outside, Washoe only heard humans talk at a late stage, as the Gardners insisted that all communication around her be done through sign language, including amongst humans. Washoe learned words (apple, yoghurt, doll, categories (fruits, candies, toys), proper nouns (Roger, Allen) and verbs. What she learned would let her ask precisely what she wanted, create social ties, express her satisfaction or sadness (the latter was particularly striking when her biological son died – her understanding of the concept connected to the word "death" pushed her into a real depression when Fouts gave her the news) or simply to picture the world (she would sign to herself while looking at pictures). What does this success tell us? That for chimpanzees, social links, the expression of emotions and playing are important things, and that if they find a means to stimulate these, they get to grips with it as much as possible (Washoe ended up teaching fragments of sign language to her adopted son).

On the opposite side, project Nim, led by Terrace, had an idea that project Washoe lacked rigor and that a more controlled protocol should be set up. As a result, Nim learned signs from ASL during sessions of imitation-repetition, often separated from the experimenters by bars, windows or one-way mirrors. He would get a reward when producing correct sequences. Nim ever only could put together short, ungrammatical and shaky sequences, and only when he wanted food. It should be noted that, strangely, the scientific community long gave priority to the failure of Terrace over the success of the Gardners, even though their project predated Nim. What does this failure tell us? That language is not a hollowed-out, emotionless, pure ability that can be learned abstractly (this point was often reproached to Terrace by Niklas Forsberg (et al., 2012)). Nim had no reason to learn a language, except to get apples, which he managed to do fairly quickly. Deprived of games and social interaction, he had no motivation to go through the substantial effort – for a chimpanzee, but also for a young child – of learning a human language. This failure, consequently, was no surprise to specialists in language acquisition in children, to whom the idea of learning a language through a vocabulary list and grammatical rules rather than though experience and stimulation made no sense.

This creates a second challenge. How are researchers supposed to study phenomena that are weakened, if not destroyed, by a controlled experimental environment? While laboratory work remains indispensible to study specific micro-phenomena, or already known elements, it is the field, even if it was semi-captivity, that allows for the observation of these events and the understanding of their operation. Experimentation time should therefore be shared between inside and outside the laboratory while giving more importance to processes that are less controlled.

This injunction may seem paradoxical. How could accuracy and precision be gained through accepting testing protocols with fewer controlled parameters? This is actually a double question. The first aspect is that it is not necessarily about removing all control factors, but to observe them and remove those that add biases

Fig. 6.1 Washoe playing with Roger Fouts by throwing a doll – Research photograph by Allen Gardner

instead of avoiding them. One previously mentioned example is the very good (and much too rare in comparative psychology) meta-analysis by Christophe Boesch (2007) on studies maintaining that a young human child was superior to a young chimpanzee. Considering the differences in protocol – the young child was often in a room that he knew or had visited before, if that was not possible some familiar items were put in the room; the test was presented as a game, and while he could not always stay on the lap of one of his parents during the test, one of them would generally remain, visibly, in the room; while the young chimpanzee was often put in a room inside a laboratory, far from his enclosure, sometimes even separated from humans by a one-way mirror to avoid all contact with the animal and any "Hans the Clever" effect (Rosenthal 1970) – it is strange, to say the least, that the comparison of the obtained results from such different conditions was deemed convincing.

Beyond the location of those tests – which clearly did not seem appropriate for young monkeys – there is also the question of what can be perceived as bias. Strangely enough, in this case, nobody seemed to think that having the young children on their parent's lap constituted a bias, but everyone would have argued that holding the young chimpanzees by the hand would have.

Rationally, the second aspect about the control parameters entails taking into account the nature of what is being studied. To take a crude but telling example, it is not possible to study mourning in a laboratory because it is not possible to get, just like that, a deep, standardised emotional attachment from one subject to others and have them die one by one from complex diseases that are different every time to observe whether the reactions are different depending on the death being violent, predictable or natural.

If, by pure and extraordinary chance, such an experiment was possible, it would still be extremely complicated to defend when facing an ethics comity – the continuators of Harlow are often criticised on this point (Despret 2009), this will be discussed in Chap. 7.

Going back to the case of emotion, my main study subject in zoosemiotics, it exists in a subject with a completely subjective feelings, so the idea of looking for standards completely goes against the very nature of the phenomenon. Similarly, language (no matter what definition we give it) develops through, and for, social interaction, and trying to remove biases cripples the learning process. Complex thought is formed when dealing with a rich, diverse environment, so a controlled environment will not allow researchers to study what they wanted to create. When studying these phenomena, the laboratory *is* a bias in and of itself. The laboratory environment simply is not relevant (in the sense of the relevance theory (Sperber and Wilson 1989)) for the subject that experiences these phenomena, and the chances that they develop normally are thin.

This is where one last challenge appears, one that is doubtlessly deeper and more problematic: our vision of what is, or is not, science. We are probably still bearing the marks of behaviourism, and as a result tend to think that a phenomenon, in order to be scientifically validated, must be perfectly reproducible in a laboratory using standardised tests. While this is a perfectly valid thought process for many things in many fields, it reaches its limits when studying things that are as complex and swaying as thought, language or emotion, none of which can be standardised without also being sterilised. This danger of not seeing that certain highly interesting, important or existing elements were being excluded from classical science due to a lack of objective, standardised study possibilities had already been condemned by Daniel Yankelovitch (Smith 1972).

I am not trying to say here that nothing interesting ever gets produced in laboratories. On the contrary, as De Waal and Robert (2002) very well explained in *De la réconciliation chez les primates*, the laboratory is a necessary step in which to check, nuance, refine and more precisely understand the mechanisms observed in the field. But it should be underlined that it cannot be the only place of acquisition of scientific knowledge, and I will go even further, and say it is not the most adequate place for a certain number of subjects. The recent discoveries of Jill Pruetz (2011) on altruism in chimpanzees with no parental connection to each other, of Helmut Prior et al. (2008) on self-awareness in magpies and of Vyacheslav Ryabov (2016) on the possibility of discrete language units in dolphins are all major discoveries, and are all the result of field work.

Of course, there will be some laboratory work to decipher the videos of Pruetz, the sound recordings of Ryabov and test the magpies of Prior with mirrors, items, and different situations. And the results from the laboratory will doubtlessly show some lines of research, some grey areas, some contradictions to previously existing conclusions that will send researchers back to the field, and so on. Therefore, the idea certainly is not to completely exclude the laboratory, but to end an implicit doctrine that says it is the only reliable place for knowledge, and that the field is simply a very unrewarding annex where we prefer to just send students.

There are a lot of things to learn from field work for those who take a more rigorous look at it, and the scientific community has a lot to gain from a bigger collaboration between field work and the laboratory. There is no question here of going too far in the opposite direction and advocate for a new doctrine of "total field work" that would be just as absurd, biased and obtuse as the previous one. On the contrary, the goal is to ask questions and take a new look at the subjects of our studies, and the methodological reflection that always comes with an interdisciplinary work seems like the perfect moment to do so. What exactly is the nature of our study subjects? What do they interact with, which other discipline can potentially bring us answers? What kind of environment will they be the least disturbed by, and in which they will be at their most "natural" and therefore the most likely to show us their true colours, and which disciplinary field has the tools to describe them? What is a bias to them in the light of their natural state? What do I, as the person studying this subject, truly believe, and how might it unwittingly influence my subjects, my experimentation or myself? What is, after all, really relevant in science?

6.2.2 Observation Is Disruption

If the main criticism that can be made about a "exclusively in laboratory" study is that it is a source of bias and subject disruption, it should be kept in mind that is also unfortunately the case for the vast majority of field observation techniques. By definition, especially when talking about the subject of behaviour, and therefore of reactions to environment, observing is a source of disruption, as it entails introducing an element (the researcher, the camera, the recorder, etc.) in an environment where it does not usually exist.

Here, the problems met by field observation and laboratories meet. The benefit is that it is therefore possible to offer solutions that can be applied to both situations (when dealing with the "observation is disruption" aspect – the other issues that were raised by the laboratory in the previous point remain valid).

The first solution that can be offered is to take from the methodologies used by researchers in human sciences, particularly social sciences, that work on the assumption that the researcher must be considered a disruptive element of the experiment, and not a neutral and objective element (not in the sense that the researcher is not neutral or not attempting to be neutral, but rather that their influence on the experimental system is not neutral).

Other than their simple physical existence (or the existence of their device) that needs to be taken into account by creating control conditions or leaving the device "blank" long enough to let the subjects accustom themselves before doing any experimentation, the researcher must also take into consideration the ideological way they influence the experimentations, but this aspect will be brought up in the next point, as I felt it required a specific attention.

Sometimes linked to the position of the experimenter, but also sometimes made with the greatest possible care regarding objectivity, experimentation, or rather its protocol, is also an aspect affected by biases, and should therefore be made as ecological as possible (in the meaning given by Christen (2011), see next point). It is a point that has been known in science for a long time: changing the realisation conditions in physics, some variations in the environment in chemistry, an irrelevant protocol in medicine, and an experiment suddenly has completely opposite results. In the case of studying living, thinking beings, and therefore complex subjects, this is all the more true. And strictly controlling all the parameters of an experiment is not necessarily the best choice when working on live subjects, who are by nature used to the imponderable. Here is what Fouts (& Rigby, 1977) said about the experiments of Premack:

> It should also be pointed out that when an experiment is so overcontrolled, it limits the possible findings to those preconceived by the human experimenter rather than explores the full mental capacities of the subject. (Fouts and Rigby 1977 p. 1049)

What are the right conditions for an experiment? The right ones are those that least disturb the subject, so as to make sure that any change can be attributed to the only variation introduced by the researcher: the experimentation subject. Here too, the input of experimentations from human sciences can be particularly interesting as those researchers are used to having to experiment with humans, to whom it is sometimes necessary to lie or ask to do several useless tasks to diminish the bias of the device during the observation.

Beyond environment is the question of relevance of the test, and how to adapt it to the test subject to get the most telling observation possible. To give a crude but telling example, it is neither useful nor convincing to test the artistic sensibility of a deaf person by making them listen to Mozart. Rather, they should be asked what they think of Picasso. When put this way, it seems obvious, yet it is far from being the norm, especially when dealing with species very different from ours (or, on a larger scale, when dealing with different intelligences, as this challenge also exists with machines (Grandgeorge et al. 2013)).

In ethology, we can, have, and will continue to mention the mirror test, with all the information it gave as well as all its limitations (Gallup 1977). The issue is not really that it has limits, as all experiments do, but rather that these limits were long interpreted as cognitive limits for the tested individual, and not limits of our possibility to observe a phenomenon through this means. The relevance theory is once more an excellent way to sort the conceptual and experimental tools so as to only keep those that are appropriate for the test subject.

Lastly, we should not forget about external biases. The very telling case of the experiments of Nissani (2006) on wood-carrying elephants in Burma can be mentioned. The elephants are first trained to lift the lid of a designated box and to take the treat inside of it. Then, they are showed several boxes, and only the one with the treat in it has its lid on the side; all they have to do is dive in with their trunk.

However, they make themselves lift the lid and put it back down on the side of the box before grabbing the treat. It could be concluded that they just "stupidly" learned a sequence, that this is not a thought process but rather just conditioning, but the bias that this type of experiment can introduce should also be considered. This type of experiment is a demand, an order. And as Yves Christen explains it, "an Asian elephant working for Man [is] a broken-down animal. […] They will do everything to respect what they think are the orders of their torturers" (Christen 2011, pp. 501–502). This is a perfect example of a strong bias (since it influences the entire experiment here) that is external to the test, but revealed by the chosen mode of experimentation. The IFC tool can prove particularly useful in this context – to what degree is the action intense, physically or psychologically, for the subject? To what degree is the subject familiar with this type of action, is it frequent? What meaning can this type of action have in the context of the life of the subject?

6.2.3 How Ideology Can Be Rooted in Science

The notion of ecology has only recently begun to be applied to the ethological field in the sphere of researching intelligence and language in other species. In his book *L'animal est-il une personne?*, the biologist and neuroscience specialist Yves

Christen (2011) often invokes this notion to explain the variations in results, which are sometimes very large, that can be found between experiments, and defines what forms, in his opinion, an "ecological", convincing test that respects a being that cannot speak.

The result of an experiment is never an abstract, autonomous entity: the experiment conditions, the test protocols can influence it, but so can the stance of the experimenter or even the simple opinion they have of the experiment. If one still needs to be convinced, they only need to read the works on the placebo effect (Chamayou 1999). A large variety in test results appears depending on whether they know the medication they are taking is a placebo, but also depending on whether the *doctor* giving them the treatment knew it was a placebo. This is the origin of the generalised "double-blind" experiment protocols that are mainly used in medicine. While it is not always possible to experiment in these conditions, and to therefore quantify the influence of the stance or belief of the experimenter, it is still possible to study the main sources of this subjectivity to more easily identify them and actively work to make their stance as neutral as possible.

It is important, first and foremost, to pay attention to the fact that an experimenter is rarely an isolated individual without any influence. They generally belong to a scientific school of thought, they agree with some ideologies and sometimes defend them. This, in itself, is not necessarily a problem; everyone more or less agrees with some ideologies and everyone belongs to a school of thought (including sometimes to the school of thought of those refusing to declare themselves as part of a school of thought). The problem arises when there is not a complete acknowledgement of this scientific bias.

Within life sciences (particularly ethology), this bias can manifest itself in a very visible way in the remarks of scientists that strongly maintain they are "against anthropomorphism" but do not recognise that they are part of another movement that can be just as problematic, anthropocentrism (Christen 2011).

This has already been discussed in Chaps 1 and 2; this position can be defended and is not necessarily a problem in and of itself, even if I do not agree with it. But, again, this ideological stance must be clearly stated and it is important that the reader of the works of these scientists knows about the potential influence of this stance on the results. It is for this reason that I clearly stated my position on the topic in Sect. 2.3.1.

Griffin (1977), who has already been quoted in this book, pointed out this tendency and insisted upon the problems it created for researchers that would leave themselves open to it:

> [...] it has seemed of fundamental importance to many scholars and philosophers to insist on a "radical difference in kind" in the terminology [...] rather than a quantitative difference in degree of elaboration. [They] have struggled to formulate objective criteria by which human language can be qualitatively distinguished from animal communication. But this task has become increasingly difficult as more and more is learned about the communication behavior of honey bees, chimpanzees, and other animals. (Griffin 1977 p. 29)

Another good example of ideological impact can be found in the handbook of *Ecologie Comportementale* edited by Dunod, which makes very clear the way the arrival of the cognitivist approach in ethology influenced the mindsets of researchers, and brought completely new notions, such as the "cognitive map" to describe the origin of a behaviour, or "insight", which had been exclusive to human psychology until then, but is now considered part of animal ethology (Danchin et al., pp. 12–13).

These latent ideologies can cause methodological problems even in the way experimentations are set up. As a result, the hypotheses and premises constitute a second type of possible biases directly connected to the ideology. Obviously, it is impossible to begin a study, experimental or not, without any hypothesis and without any premise, but these, too, should be called into question a lot more than they have so far, at the risk of creating results that are skewed by expectations that were too precise or premises that were too narrow.

Within ethology, there have been a certain number of researchers that would simply refuse the findings of their colleagues that tended to prove the presence of emotions, intelligence, or language in other animal species and do their own experimentations, which would turn out to be failures that could be directly attributed to the chosen methodology. The case of Terrace (1987) doing so with the works of the Gardner and Gardner (1969) or Premack and Premack (1972) is a perfect example.

This kind of stance cannot be credibly held for two reasons. The first is connected to life sciences, because this position calls into question the very foundations of its methods (similar results being found by different teams, at different times, with varying methods and subjects are valid results). The second is more of a human sciences problem, since researchers, like philosophers, often struggle with providing a precise definition of the problematic concepts (this was detailed at length in Chap. 3). Proving they are exclusive to humans leads to series of logical and ethical absurdities (the case of Guyomarc'h (1980, p. 168) comes to mind – he would go as far as putting humans that could speak on one side and humans that could not speak with the *Pongidae* on the other to keep spoken language as a human characteristic, which is a process I find dubious, both methodologically and ethically).

Lastly, it is important to keep in mind that any researcher, far from being a neutral and objective entity, is a human with a history (both on a professional and personal level) which is a source of influence. The different histories, especially in interdisciplinary work, are an undeniable enrichment as long as they are, once again, taken into consideration, rather than dismissed as not having any possible influence.

Of course, these biases can show themselves in other places than just life sciences, but they happen to be particularly present due to the sentience of the subjects. In a similar fashion, a still widespread opinion in the study of behaviour is that only the ethologist, the scientist, strictly speaking, is authorised to study and understand animals and bring proof of the existence of this or that aspect of a behaviour. This stance has long deprived studies of data from veterinarians (Béata 2009; Passelegue 2009), carers or pet owners. The view of academic knowledge being above knowledge from the field has only started to fade out very recently – allowing me to include new documents in the corpus of my thesis, see Sect. 1.3.2, which would have been unthinkable just a few years ago.

Also in ethology, the tradition of comparing results of animal studies with human studies presents three biases that are quite fragile and not very desirable for a researcher trying to keep the ecology of their experimentation: generalisation, references to closed categories, and an incorrect distinction between real and potential abilities.

The case of generalisations consists in considering one or several specific cases to be indicators of a general rule. In ethology, we talk about monkeys having learned sign language by using their names, because they are special cases and most primates do not communicate that way. Nonetheless, as more and more subjects have succeeded in learning sign language, scientists have concluded that, while there may be exceptions, all chimpanzees and gorillas can potentially learn sign language. This is a classic generalisation process.

The problem occurs when it is not respected. Hence why Terrace (1987) claimed he could prove the total impossibility of chimpanzees mastering sign language by using the failure of his one project, project Nim (and the fact that it is disputed by other researchers has already been mentioned). This was the case for talking birds for a long time, as it was often said that they could very well repeat but could absolutely not understand what they were saying and what they were being told. We had to wait until Pepperberg (1999) studied the question and changed the proposed way or teaching (by showing what to do with an assistant, rather than asking the bird) for the experiments of project Alex to show that parrots are perfectly able to understand what they are told and what they say, if taught with a method that is different from the one that had been used until then.

The references to closed categories of population are often found in comparisons of ethological studies with humans, even without us realising it, simply by taking a specific category of human beings and thinking it represents all of them.

This particularity was first brought to light by anthropologists specialising in some so-called primitive tribes, within which they would discover that what we considered to be common abilities that all humans had were not always present, or at least not in the expected way. The case of the Pirahãs, an isolated tribe in the Amazon rainforest, a case that was very much in the public eye, was a particularly interesting over, as their ability to count, to compare ("bigger than", "smaller than") and to talk about what had happened a long time ago have been asserted as being inferior to the same, naturally present, abilities in great apes, and the gap widens even more if we bring up the primates that were acculturated by experimentation. While the works on this topic can be debated (Velmezova 2018), they had the merit of shining a light on a bad habit of life sciences in these comparative studies. It turned out that most comparisons with humans defined "humans" as "average human in modern western society", which is not exactly the same thing.

This idea of categories raises a lot of questions on what actually is the norm, even in our species. Because while there is a biological norm (a child that is deaf, dysphasic, mentally retarded, etc., is, from the viewpoint of the biology of our species, "abnormal"), the notion of norm is more of a social question – it was not that long ago that it was considered abnormal to be left-handed, and children were being taught to write with their right hand, which was, paradoxically, biologically abnormal for them and created cases of dysgraphia.

The last argument I will develop is the difference between real abilities and potential abilities when we do a comparison of those between humans and animals. Real abilities are the ones an individual uses effectively, potential abilities are those that they could have developed, but do not have.

The example of bats is often mentioned; bats have this perfected radar that humans, because they do not have it from birth, built by creating the sonar. However, this is incorrect: a *tiny minority* of humans can build it, just like the building of planes replaced flight, building ships replaced swimming, etc. These are potential abilities, but a crushingly large majority of humans do not possess them as real abilities. This distinction, which seems to barely be taken into account in ethology, is very well understood by the social sciences and language sciences (especially in sociolinguistics), and this is another situation where the meeting of methodologies can help solve problems in other subjects. Thus, particularly talented subjects amongst animals, particularly great apes, have often been deprecated by underlining the gap between them and our biggest geniuses: Einstein, Mozart…while forgetting the gap between these great minds and the vast majority of our species is also gigantic. A more objective approach of this comparison can be provided:

> *Il serait faux de nier ses différences fondamentales que représentent pour l'homme une culture hyper développée, notre capacité de raisonnement et le potentiel démesuré que nous avons de transformer notre environnement de manière à influencer notre propre avenir. Nous prétendons que ces traits distinctifs, et il en existe d'autres, ne sont pas des discontinuités entre nous et les animaux mais des différences d'ordre quantitatif ou de degré.* (Danchin et al. 2005 p. 523)
>
> Translation:
>
> It would be wrong to deny the existence of these fundamental differences represent a hyper-developed culture for humans, our reasoning capacity, and the immeasurable potential that we have of transforming our environment to influence our own future. We predicate that these distinctive traits – and there are others – are not discontinuities between us and animals, but rather quantitative differences or a question of degree.

References

Alexeev, A., et al. (Eds.). (2017). *Questions Actuelles de la Neurophilosophie* (translation from Russian title). Moscou: IINTELL.

Béata, C. (Ed.). (2009). *Zoopsychiatrie : L'attachement*. Marseille: Solal Editeurs.

Bekoff, M. (2009). *Les émotions des animaux*. Paris: Éditions Payot & Rivages.

Bertier, J. (1994). *Histoire des animaux : Traduction, présentation et notes*. Paris: Gallimard.

Bickle, J. (Ed.). (2009). *The Oxford handbook of philosophy and neuroscience*. Oxford: Oxford university press.

Bodamer, M., & Gardner, A. (2002). How cross-fostered chimpanzees (*Pan troglodytes*) initiate and maintain conversations. *Journal of Comparative Psychology, 116*(1), 12–26.

Boesch, C. (2007). What makes us human (*Homo sapiens*)? The challenge of cognitive cross-species comparison. *Journal of Comparative Psychology, 121*, 227–240.
Bradshaw, I. G. (2004). Not by bread alone: Symbolic loss, trauma, and recovery in elephant communities. *Society and Animals, 12*(2), 143–158.
Chamayou, G. (1999). L'essai "contre placebo" et le charlatanisme. *Pour la science, 38*, 14–17.
Christen, Y. (2011). *L'Animal est-il une personne ?* Barcelone: Flammarion.
Danchin, E., Giraldeau, L.-A., & Cézilly, F. (Eds.). (2005). *Écologie comportementale: cours et questions de réflexion*. Paris: Dunod.
Darwin, C. (1872). *The expression of the emotions in man and animals*. Londres: John Murray.
de Fontenay, E. (1998). *Le Silence des bêtes*. Lonrai: Fayard.
De Waal, F. (2013). *Le Bonobo, Dieu et nous*. Lonrai: Les Liens qui Libèrent.
De Waal, F., & Robert, M. (2002). *De la réconciliation chez les primates*. Paris: Flammarion.
Delahaye, P. (2017). Cognition, langage, émotion : faut-il sortir du laboratoire ? In A. Alexeev et al. (Eds.), *Questions Actuelles de la Neurophilosophie* (translation from Russian title). IINTELL: Moscou.
Despret, V. (2009). Il est temps d'en finir avec Harlow. In C. Béata (Ed.), *Zoopsychiatrie : L'attachement* (p. 13). Marseille: Solal Editeurs.
Forsberg, N., Burley, M., & Hämäläinen, N. (Eds.). (2012). *Language, ethics and animal life*. New York: Bloomsbury.
Fouts, R., & Mills, S. T. (1998). *L'école des chimpanzés: ce que les chimpanzés nous apprennent sur l'humanité*. Paris: J. C. Lattès.
Fouts, R., & Rigby, R. (1977). In T. Sebeok (Ed.), *How animals communicate* (p. 1049). Bloomington: Indiana University Press.
Freeberg, T. M. (2012). Geographic variation in note composition and use of *chick-a-dee* calls of Carolina Chickadees (*Poecile carolinensis*). *Ethology, 118*, 555–565.
Gallup, G. G. (1977). Self-recognition in primates: A comparative approach to the bidirectional properties of consciousness. *American Psychologist, 32*, 329–338.
Gardner, A. R., & Gardner, B. T. (1969). Teaching sign language to a chimpanzee. *Science, 165*(3894), 664–672.
Grandgeorge, M., Le Pévédic, B., & Pugnières-Saavedra, F. (2013). *Interaction et Intercompréhension : une approche comparative homme-homme, animal-homme-machine, homme-machine*. Bruxelles: E.M.E.
Griffin, D. (1977). In T. Sebeok (Ed.), *How animals communicate* (p. 29). Bloomington: Indiana University Press.
Guyomarc'h, J.-C. (1980). *Abrégé d'éthologie* (p. 1980). Paris: Masson.
Lorenz, K. (1967). *Évolution et Modification du comportement: l'inné et l'acquis*. Paris: Payot.
Machamer, P. (2009). Learning, neuroscience, and the return of behaviorism. In J. Bickle (Ed.), *The Oxford handbook of philosophy and neuroscience* (p. 175). Oxford: Oxford University Press.
Morin, E. (Ed.). (1974). *L'Unité de l'homme : invariants biologiques et universaux culturels*. Paris: Éditions du Seuil.
Moss, C. (1989). *La Longue Marche des éléphants*. Paris: Robert Laffont.
Nissani, M. (2006). Do Asian elephants (*Elephasmaximus*) apply causal reasoning to tool-use tasks? *Journal of Experimental Psychology: Animal Behavior Processes, 32*(1), 91–96.
Passelegue, P. (2009). Épidémiologie des hyperattachements. In C. Béata (Ed.), *Zoopsychiatrie : L'attachement* (pp. 71–76). Solal Éditeurs: Marseille.
Pepperberg, I. (1999). *Alex studies. Cognitive and communicative abilities of grey parrot*. Cambridge: Harvard University Press.
Premack, A. J., & Premack, D. (1972). Teaching language to an ape. *Scientific American, 227*(4), 92–99.
Prior, H., Schwarz, A., & Güntürkün, O. (2008). Mirror-induced behavior in the magpie (Pica pica): Evidence of self-recognition. *PLoS Biology, 6*, e202. https://doi.org/10.1371/journal.pbio.0060202.

Pruetz, J. D. (2011). Targeted helping by a wild adolescent male chimpanzee (Pan troglodytes verus): Evidence for empathy? *Journal of Ethology, 29*, 365–368.
Rattasepp, S. (2018). *The human mirror. A critique of the philosophical discourse on animals from the position of multispecies semiotics*. Tartu: University of Tartu Press.
Rosenthal, R. (1970). Le préjugé du maître et l'apprentissage de l'élève. *Revue française de pédagogie, 13*, 39.
Ryabov, V. A. (2016). The study of acoustic signals and the supposed spoken language of the dolphin. *St. Petersburg Polytechnical University Journal: Physics and Mathematics, 2*(3), 231–239.
Sebeok, T. A. (1976). *Studies in semiotics: Contributions to the doctrine of signs*. Bloomington: Indiana university press.
Sebeok, T. (Ed.). (1977). *How animals communicate*. Bloomington: Indiana University Press.
Smith, A. (Yankelovich, D.). (1972). *Supermoney*. New York: Popular Library.
Sperber, D., & Wilson, D. (1989). *La pertinence: communication et cognition*. Paris: les Éditions de Minuit.
Terrace, H. S. (1987). *Nim*. New York: Columbia University.
Velmezova, E. (2018). About the (semiotic) limits of the human language: Discussing the case of Pirahã. *Sign Systems Studies, 46*(2/3), 392–397.
von Uexküll, J. (1934). *Mondes animaux et monde humain. Théorie de la signification*. Paris: Denoël.

Chapter 7
Animal Studies, Animal Ethics

Abstract Evaluation sub-question n°1: what are the problematic ethical aspects of animal studies that semiotics can answer?

This chapter is meant to present the different problems one can be confronted with while studying animals, starting with general issues. It will thus tackle the philosophical difficulty represented by the fact of working with living beings, especially the difficulty in trying to understand foreign minds. The question of emergency situations and their handling will also be addressed. It will then examine the issues regarding ethics applied to animal studies directly, such as the problems raised by the study of endangered species, cognitively complex species, or even the question of animal suffering. In the end, we will present solutions that the semiotic methodology brings, both for general issues and for more specifically ethical problems.

Keywords Animal ethics · Endangered species · Pain · Cognitive abilities · Population management

In this chapter, we will bring answers to the first evaluation sub-question from the introduction: What are the problematic ethical aspects of animal studies that semiotics can answer?

Firstly, it seemed interesting to return to the more general philosophical problems that can be encountered while working on animals. These problems are not ethical problems, strictly speaking, but I thought they could cause researchers to face certain dilemmas, and therefore should be discussed.

Thus, we will talk about the challenges of working with living beings, who, by nature, vary from one individual to the next and can only partially provide a source of reproducible experimentations (this point was already mentioned in Chap. 6). We will particularly focus on the difficulties of working on and with fundamentally foreign minds that can only be imperfectly understood. Following this, we will discuss the fact that life sciences are currently facing an emergency regarding their study subjects, which leads to numerous problems and sometimes non-optimal solutions.

We will then turn our attention to animal ethics issues specifically.

P. Delahaye, *A Semiotic Methodology for Animal Studies*, Biosemiotics 19,
https://doi.org/10.1007/978-3-030-28813-6_7

To answer the last point of the previous part, this chapter will begin by discussing the case of working on endangered species and the unique problems this situation creates. I will also bring up a topic that was a part of my thesis – problems that come with the study of complex species, and in what way ethics are particularly important in these cases. Finally, the discussion will turn towards the currently quite controversial topic of animal suffering, whether caused by science or proven by science.

The last part of this chapter will be devoted to the semiotic viewpoint on these different problems and the solutions the discipline can bring.

That part is split between semiotic solutions to general animal study problems, and solutions to more specific ethical problems.

7.1 Issues in Animal Studies

Talking about animal studies entails sooner or later, especially in the current societal context, talking about the issues that stem from studying species that are fundamentally foreign to us. Not all of these issues are necessarily ethical, however, some of them are, for example, philosophical ponderings, dilemmas or conception problems, and I think it interesting to begin the chapter by studying them.

Therefore, this part will focus on issues met while working on my thesis. Some had the opportunity to be discussed with members of the jury, some of whom were from the world of semiotics and others from life sciences – as a result, they integrated particularly well the interdisciplinary aspect in their study.

Logically, the first question that comes to mind is about the expectations and limits of working with living beings, especially animals. What can be learned from them? What cannot be learned? What are the expectations of this type of study, what are we really trying to understand? What are the methodological, practical, ethical limits (even if this last one will be discussed in the second part of the chapter)?

Then comes the obvious follow-up question, one that is dear to Uexküll, of working with minds that are fundamentally different from ours, which has its own behaviour and semiosis (Kull 1998). This philosophical question of what is and what is not comprehensible or knowable in another, fundamentally foreign individual will be discussed at length. This point will allow establishing what I consider to be a "final frontier" of knowledge in animal studies, but also to explain why this is not a problem in itself (the solutions to take this into account and adapt to it will be detailed in Sect. 7.3.1).

Lastly, in the light of the current state of biodiversity, the primordial backdrop of life sciences, I thought it necessary to discuss the issue of dilemmas that can face researchers because of the emergency they find themselves involved in despite themselves. What or who should be studied when both time and means are lacking? Should we prioritise the most useful species, the most endangered species, the species closest to us? Should we get involved and intervene, at the cost of distancing ourselves from the alleged scientific neutrality and non-intervention?

7.1.1 *Working with Living Beings*

Working with living beings, particularly beings from the animal kingdom, raises specific issues that do not appear when working with formulas or molecules. Some of these issues are ethical – those will be discussed in the next part of the chapter. Some are statistical issues: there are many more variations between two individuals of the same species (especially if the species is particularly cognitively complex) than between two molecules of the same substance, and this makes perfect sense. This also entails that we should think differently about the methods that let us learn something, and keep in mind that there are always things that we miss (especially in the mental sphere, this will be discussed in the next point).

Working with living beings is first and foremost working with species that are foreign to us, with which some comparisons, some approximations are possible, and with which, on the opposite side, we need a lot of imagination and rigor to understand how they perceive a radically different world from the one we perceive, and how they evolve in it.

The importance of a good understanding of the way an individual's senses work to comprehend their mental world has already been mentioned in this book. It is important to find out which senses are the main ones, the secondary ones, and which ones are sign carriers, and how; there are a number of questions necessary to obtain a good analysis that is not just semiotic but also ethological.

While this is a problem that seems to have already been solved by the advances made in biology, there is actually still much for us to understand, discover, and maybe even rediscover.

A recent study in cetology agrees with this assessment. I will detail the study here to extract from it what I believe to be causes for reflection. Recently published in the *St Petersburg Polytechnical University Journal: Physics and Mathematics*, it is the work of cetologist Vyacheslac Ryabov (2016), who has been working for years in the Karadag reservation in Crimea. Until then, we knew about the echolocation functions of ultrasound frequencies emitted by dolphins. We also knew that they had a communication system based on postures and "clicks", but also suspected that we were missing some things considering their great capabilities when it comes to understand human instructions. Particularly, their sensitivity to "grammar" (the different interpretation of signs in the ABC combination, BAC combination or CAB combination) raised questions.

Ryabov brings innovation with two points. The first was to postulate that when we try to understand dolphins, to reconstruct the mental map of what their different sense represent for them, we do not have all the required elements to do so. Some senses, or at least some levels in these senses, are missing. As a result, he broadened the range of his recordings, by measuring, for example, exchanges through ultrasounds between individuals going all the way to 200 kHz, while most of the research on communication in cetaceans focuses around the 20 kHz range.

The second point was to postulate the existence of a very advanced and elaborate form of communication, since sensitivity to grammar had been noted, that could potentially use the same channel as echolocation (in the same way that our voice is used to form coherent sentences, but also to scream in pain or anger).

The works of Ryabov note several things:

- When measured at this frequency, the pulses produced by dolphins form complex groups in packs.
- Some of these pulses, which he names NP for "non coherent pulses", are emitted with much smaller intervals than others.
- These NPs are exclusively produced during interactions between dolphins, even when they are not moving, and never while searching for food (as we know, their search for food uses echolocation)
- These productions are "conversation" productions, meaning that the production from one individual never overlaps with one from another.
- To Ryabov (this is still just a theory), NPs could be words or phonemes, perfectly identifiable and distinct from each other, with their own characteristics (amplitude, frequency, delay), and the packs of NP could be sentences.

This study raises a number of questions in an interdisciplinary work for several reasons, both biologic and semiotic. I believe this type of question can be applied to other studies on other species, since it is about dealing with the very difficulty of trying to understand living beings.

The first challenge is that, despite the current state of knowledge, it is still possible that we are missing out on part of the senses of an animal, and therefore part of the signs and part of the meaning. Thus, when we notice the existence of a complex ability (in the case of dolphins, the sensitivity to grammar), we should systematically consider researching it: abilities of this type require meaning, and the meaning requires signs – so signs must exist somewhere and just cannot be seen yet.

The second is that as humans that possess sensory faculties, and therefore are limited by them, we need to show resourcefulness and technique to reach senses that are not ours. Ryabov himself says it is no surprise we still struggle to understand the language of dolphins: decrypting the recordings he produced requires recorders with very broad ranges, able to analyse minuscule variations in pressure or time gaps that can be measured in milliseconds. The question of main senses is therefore not just about ecology and adapting to the individual as we know them, but also a hypothesis regarding the way to work when encountering phenomena that have been observed or documented but that remain unexplained on a semiotic level: "something is producing meaning, but how do we find it and observe it"?

7.1.2 *Difficulty to Understand Stranger Minds*

Using the points raised by the previous section, I would like to keep helping this research grow by providing an avenue of reflection which is mostly philosophical but also possesses obvious practical applications. This is valid for different types of studies that aim to approach and understand animals within their "black box" (Eco

1999). In my case, it took the shape of a question that focused more on emotions that I believe constitutes a good example: what is a biological or semiotic work on emotions?

While I do not have the pretention of giving a complete, absolute response to this question but rather give an answer that seems correct, in accordance with the possibilities, limitations and opportunities that I see in this research field. During the entirety of the following point, I will be talking about emotions because it is my area of specialty, but each occurrence of the term "emotion" could easily be replaced by "conscience", or "cognition" or other terms.

A research work on emotions (or at least for the disciplines that are relevant to my work and in which I worked during this research) is an attempt to study in an objective, reliable and relevant way an exclusively subjective phenomenon which only exists within an experience of an individual. This is also the case when studying animal cognition, conscience, long-term memory or any other internal phenomenon. I believe it is an extremely difficult endeavour, but also an extremely necessary one. As Yankelovitch (Smith 1972) would say, just because something is difficult to measure or problematic to evaluate does not mean it is not important or does not really exist. His exact quote was, to think this way is, scientifically speaking, suicide.

What are we really studying when we are studying subjectivity? Again, looking at the emotions' case, I would like to say that we study emotions "the way they are felt" and not "the way they are". Because there is no way to know objectively "the way they are", but there are plenty of tools and methodology we can use to understand how a subject lives and feels an emotion. Ethograms can be used to notice a change in habits or behaviour. Long term studies can observe the impact of a particularly emotional event (a death, trauma or birth) on an individual life. Some pathological behaviours such as prostration, refusing to eat, or carrying a dead baby, can alert us on the impact some of these events can have on subjects and how an emotion can fill their "mental world" (or Umwelt, if we want to use von Uexküll's terminology) so much that it stops them from living. There is a lot of semiotic tools that can be imported into the biology field in order to study emotions *in vivo*, not by an objective definition but by the semiosis of subject experimenting emotions.

For another point, I would like to underline that a semiotics of emotions (or self-awareness, or memory, again) must be, by deduction, a flexible model, able to take note of the species of subjects, their family group, their personal background and their personality. It's, indeed, complex work, but in no way impossible. In fact, we have been doing so with our own species for a while, in sociology, psychology or medicine. By doing so, the semiotics of emotions can study a real kind of emotion – the one which is felt by the subject creating it – and not an ideal but hollow view of what emotion should be.

At this moment, another common question appears "how can we be sure that what we call sorrow in other animals is the same as the one we feel as human beings?". This question, or to be more precise, the different variations of this question, was historically used to disregard a lot of subjects when it came to discussing other animals. Are their tools the same as our tools? Is their self-awareness the same as ours? How can we be sure that when gorillas or chimps use signs language, they use it the same way we use it? They were often used to claim that the subject was without scientific interest, childish, and that no serious scientist should lose his or her time with this. And all this time, there still were scientists to observe chimps making kapok-proof shoes to climb trees, to teach gestural syntax to dolphins or to put magpies in front of a mirror and make important discoveries. And now, we have this one: how can we be sure that our emotions are the same as what seems to be their emotions? My answer is very simple. We cannot. We cannot because, even in our species, this is a frontier we cannot cross, what I call a "final frontier". Why would emotions be a final frontier for the sciences of live? Let's take a short example, inside our own species, and about what is probably the most debated emotion of all time: love. If I love my husband, I have no way to be sure that what he feels, and he calls love is the same as what I feel. I cannot know, and I will never be able to do so. If I put him in a scanner, I might be able to see that the same neurons shine when he thinks about me than when it is me in the scanner thinking about him. But that would not tell me how he feels.

That's the moment where semiotics of emotions enables us to keep working. Because, if we have no way to know if it is really the same to be jealous when you are a human being or when you are another human being or when you are a gorilla, but if the signs, the consequences (on social life, for example), or the behaviour are the same, what would be exactly the reason not to study these emotional signs the same way? That question, "is it exactly the same?", is a pointless question. It is not that we have no answer yet, it's that there is no possible answer. This is what I call a final frontier in science: where "not yet" becomes "not ever".

But I would like, now, to underline the fact that a final frontier, and in my case this final frontier, is a particularity, not a problem. This point is only a problem if we, as scientists, make it a problem: it's not the first time that we have to study something where a part of the subject is unknowable. I will take an example in another scientific field. In physics, scientists have to study the electron. With the electron, a part of its "behaviour" is called the spin. We cannot understand what it is. And I'm not the one telling you that, that's Richard Feynman's words (Feynman 2000). But what do physicists do with this thing that's "out of human's mind reach"? They do not leave the electron alone and say "we cannot study it", they work on it and they say "everything is going on as if the spin was the electron spinning around itself". It is not what the spin really is, of course, but it is the closest we will ever be, and, what is interesting, it is close enough to make functional models and equations and to study it. That is what I would like to propose with semiotics of emotions. I do not know if my sorrow is the same as an elephant's, but "everything is going on as if" it was the same. What I know is that this sentence shows quite well the closest we will ever be able to be. So let's build a model to study it as if it was the same, and let's see if it's functional. There is not really another way to work. We should not be afraid of a final frontier. What is important is to, first, acknowledge it, and second, find a way to deal with it and continue to work.

It is important to do so because emotions may be a final frontier, but they remain an important part in a lot of lives. This subject has been put aside and disregarded for a long time, partially because of the behaviorist theories. But more recent studies tend to show how, on the contrary, emotions are a major part of life. Today, if emotions are a subject still understudied, it is not because it is a minor subject. It is much more because we have difficulty to apprehending it; we do not know how to find a good methodology, so we are tempted to just leave it alone in a corner and go back to other subjects with a lot of variables more easily measurable. But this is a dangerous scientific position to have.

7.1.3 Situation of Emergency

There are also situations that can be problematic simply because they are emergencies for which no optimal solution exists. During our presentation for the *Hybrid Natures* colloquium in Tartu in 2018, we quoted several examples that we would like to mention here to illustrate our point and the frequency of this type of event.[1]

For the animal, in an emergency situation, benevolent acts from human beings often become violence, because they carry the semiosis of violence. We must understand that, from a semiotic point of view, the animal has no way to tell apart the benevolent violence from the ordinary one. That means benevolent human actions may carry the semiosis of threat.

This semiotic perspective can have direct and important consequences. When interacting with wild animals, these consequences are maybe the most impressive and problematic. For example, in African sanctuaries, the hippos' case is a delicate matter. Whenever people need to displace an individual, take biological samples, perform small medical exams or important surgery, the animal's reaction can endanger its life. Contrary to most of the mammals, hippos feel safe in water, are good swimmers with few aquatic predators, and are not afraid of hiding under the water's surface. For all these reasons, hippos often escape to water ponds when they feel in danger. If the danger is a researcher with anaesthetic gun, the hippo will be shot, will identify human action as a predation action, will flee to water, and will eventually die of drowning after the anaesthetics kick in.

Other cases often occur when trying to free a wild animal from a trap or – and unfortunately this situation occurs more and more – from plastic waste, in which the animal trapped itself. As it is already trapped, and therefore vulnerable, the animal is usually anxious, and human action, by pulling or pushing it, touching and interacting with a hurt part of the body or simply approaching when any escape is impossible, carries all signals of mortal threat. By defending its allegedly endangered life, the individual can badly hurt itself, and sometimes even hurt the people trying to rescue it.

This semiotic misunderstanding can also be problematic, in a very different way, but yet an emergency situation, in the case of pets. When the health or life of the pet

[1] P. Delahaye, "To Care, to Look After, to Protect: Semiotics of Benevolent Violences", *Hybrid Natures*, Tartu, 2018.

is in danger, stress can be fatal, and if the only place where a pet can be helped – by surgery, exams, medicine – is stressful for it, we can have a serious and unsolvable problem.

I will briefly describe a last case that sometimes happens in reproduction programs which entail having to displace individuals quite often. These displacements are done to avoid inbreeding issues in species that are already endangered and need strong individuals to save their kind. These displacements are done by necessity and aim to obtain a healthier population. But in some species, they can be counterproductive. If individuals of the species have long-term memory and strong affective bounds, moving an individual is a loss. The new group member will be seen as a stranger for a while and is a source of stress for the rest of the group. In this situation, it would not be surprising that the birth rate drops significantly. This is the situation for elephants in captivity, for example.

When it comes to human interactions with animals, we often hear that "the end justifies the means", but what if the means are what cause the end to fail, and sometimes, to fail badly?

To explain more in detail how a semiotic misunderstanding can cause a benevolent action to fail, I will study two cases. The first one is the case of veterinary cares for pets. By pets, I mean more precisely animals that human beings keep for company, for the benefit of their social interaction. The same analysis can also apply to farm animals, like cows or pigs, to working animals, like assistance dogs, or to hybrid categories, like rabbits and horses. But I will talk more precisely about pets in the first given meaning.

When I worked with Lyon 1 University, which has a veterinary school and a quite rare veterinary psychiatry diploma, a question kept coming over and over. How can I take care of an animal who believes I am his or her enemy? If, in this situation, I use – as these veterinary students used – pronouns "his" and "her" instead of "it", it is because the first element of answer was to consider animals as having different personalities, and that this kind of problem does not occur with every animal, so we cannot find a correct solution without taking their personality into account. As it does for humans, personalities are interpretation keys for semiosis. If I am a confident person, I will not interpret a sign as a threat like I do if I am an anxious person. The first step was to acknowledge that some signs are part of violent semiosis for some individuals, whereas other individuals can be indifferent to them.

The second element of answer was a semiotic history of these problematic responses to veterinary interactions. Most of the time, they did not occur at the first visit, the first contact: the puppy or kitten is exploring, quite confident and it is a period of his or her life with a lot of change anyway. Problematic answers started later, but not for all animals. We had the hypothesis that, in the previous interactions, a part of the semiosis between the animal and the veterinary was violent without counter semiosis to calm the animal, who, therefore, memorized this interaction as a violent one. The consequences can be dramatic. A badly injured dog was for example so stressed and defensive that the veterinary could not help him until he lost so much blood that he fainted. A cat could not eat due to important dental plaque, but he was so aggressive that descaling had to be done under general anaesthesia; he was old, the anaesthesia impaired his kidneys and he died few months later of kidney failure.

Of course, emergencies regarding pets are not the same as emergencies for wild animals, and that is why I will now study the second case: the reproduction programs for endangered species. These programs can take various shapes, from the ones with an important human presence – captivity with artificial inseminations – to the ones with their organized absence – wildlife park with human protection. But they all seem to face the same kind of problems, and a part of them are semiotic problems.

A common part – the one that can be problematic – is the displacement of individuals to avoid inbreeding. This point is particularly problematic because it cannot really be avoided: the principle of being an endangered species is that there are few individuals remaining. The survival of the species implies to give birth a lot of new individuals, but a lot of healthy ones, able to reproduce, to raise other healthy younglings, and sometimes to go back to the wilderness after a long time in captivity. Inbreeding is a real danger for an already fragile population.

But with these collective considerations, the point of view of the individual is poorly taken into account. Yet, it is the individual which reproduces, and if the collective consideration does not respect its own Umwelt, the program can easily fail. For some species, displacement of individuals is not a real problem and, after an adaptation gap, the animal is totally able to reproduce normally. But in some others, an individual is clearly identified, it exists in other individuals' Umwelt, an Umwelt with life-long memory to remember it, it is not interchangeable, displacing it is breaking a bound and creating social and emotional distress.

These two cases illustrate the variety of situations that can be considered as emergencies. Whether it is dealing with animals that are sick, wounded, in danger or endangered, these situations leave very little leeway for researchers, carers and any and all humans involved in these benevolent but problematic relationships with animals. The solutions we will discuss in Sect. 7.3.2 are not optimal solutions either, but they do aim to diminish the consequences of this type of situation, on research, the humans involved in the research and the animals that are subjected to it.

7.2 Ethical Issues

We will now turn to the ethical problems faced by researchers doing animal work. By "animal work", we mean the research work with animals as its main study subject, and not cases in which animals are a means, a way to support the study of something else, as it can be the case in medicine or pharmacology. While these fields can also have ethical problems, they will only be lightly discussed in the last point of this part.

We will begin by talking about the case of works on endangered species. This point directly echoes Sect. 7.1.3., which deals with the urgent characteristic of some types of animal studies. Certain questions raised in that section have led to ethical questions. These are fairly recent, as the awareness of the danger that hangs over the concept of biodiversity is not old compared to the age of the discipline itself. Even though the danger looming over many species is a more and more pressing matter, it struggles to influence and change the ethics and practices that were historically inherited by the discipline.

Amongst these endangered species, some are particularly complex, cognitively speaking.. It would not be an exaggeration to say that all the highly complex species (great apes, cetaceans, elephants, etc.) are facing an imminent threat. If we add to this the challenges that come with studying species whose abilities, behaviours and expressions that are often very close to ours, it is easy to explain why we think these species, although they are an extremely small minority, strongly require to be handled separately in this part of the chapter, and arouse issues and ethical challenges that are specific to them.

Lastly, we will discuss the more general question of the problem of pain in animals, both when it is proven by science (and the ethical and moral consequences that these discoveries can have on society as a whole) and when it is created and organised by science (in the case of many study protocols – whether it is their point or not). This ethical question is a particularly sensitive one these days from a societal perspective, and has led to important modifications of the restraints imposed upon experimental protocols, as well as put into question certain research methods, so we have decided to take a closer look at it.

7.2.1 About Endangered Species

Amongst the emergencies mentioned in Sect. 7.1.3., we discussed species that are the subject of reproduction programs, as well as (although more briefly) the case of animals needing medical treatment or tests inside reservations. While these two situations do not necessarily overlap (not all animals within a reservation are subjected to reproduction programs), they do both belong to one specific group: endangered species that have an active human intervention attempting to preserve them.

These species can cause several ethical issues due to their particularly fragile status: observation issues, prioritisation issues and philosophical or moral issues.

First, regarding the observation issues, we have already established that it is impossible to observe without disturbing the subjects (see Sect. 6.2.2.), and that while minor disturbances often have no consequences, precise, thorough observations will always create bigger disturbances.

Not all of these disturbances are directly harmful; they can sometimes just be behavioural changes that make the observed species more accustomed to the observing species. But even these can prove to be harmful – if we take into account that poaching is nowadays one of the causes of the accelerating extinction of species, we can understand that for endangered species, fearing and fleeing from humans is essential to their survival.

But when it comes to endangered species, it makes sense that it is impossible to protect them if we do not know them, as we risk committing grave mistakes otherwise, as with the hippopotamuses from Sect. 7.1.3, or the recent case of the black rhinoceroses in Kenya.[2] This creates a dilemma for researchers: should they give up on observation (or at least some of them), at the risk of not understanding and protecting properly, or should they study the species as much as possible, at the risk of disrupting the species and hasten their decline? The more endangered the species, the more pressing this question becomes (and there is, on top of this, the scientific risk of definitively losing any chance of gathering knowledge on a species if it ends up extinct).

The problem becomes all the more urgent as the number of critically endangered species increases (see Fig. 7.1 – distribution of species by degree of endangerment). A large number of these species and complex species, meaning there is still a lot to learn about them (see Fig. 7.2 – distribution of endangerment per type of species).

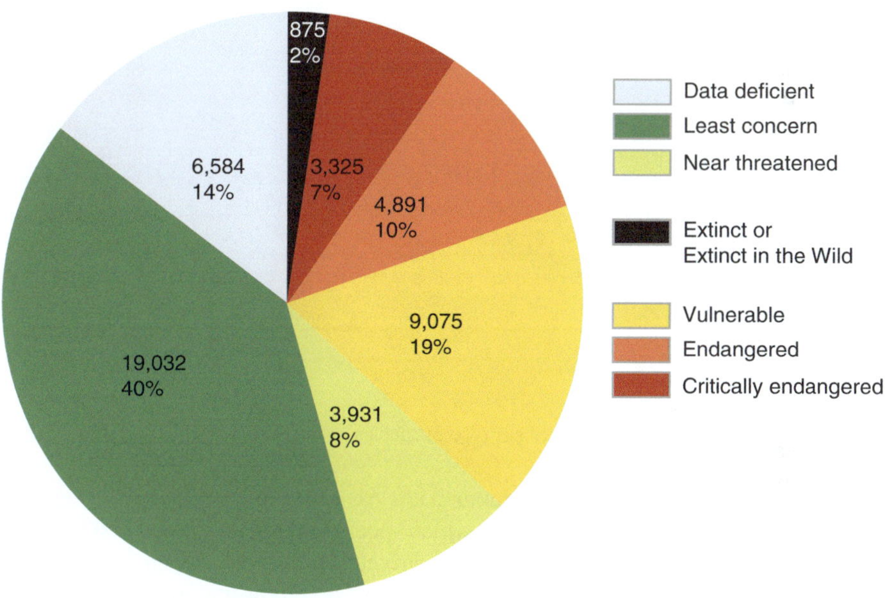

Fig. 7.1 Distribution of species by degree of endangerment. (Source: IUCN)

[2] On June 26th, 2018, 11 black rhinoceroses are moved from the Nairobi park to the Nakuru park by the Kenyan Wildlife Service (KWS). The ignorance of the reactions of the rhinoceroses led to a series of mistakes (the cages were too large, unsuitable sedation, inadequate surveillance) which caused a grave stress syndrome. This, coupled with the stifling of warnings regarding the high salinity of the watering hole for the animals, who ended up suffered from salt poisoning, resulted in the deaths of 10 of the rhinoceroses. Alive but weakened, the eleventh was unable to properly defend himself against an attack by lions days after his transfer, and he died from his wounds on July 31st, 2018. The story has been described as the biggest fiasco in wildlife conservation.

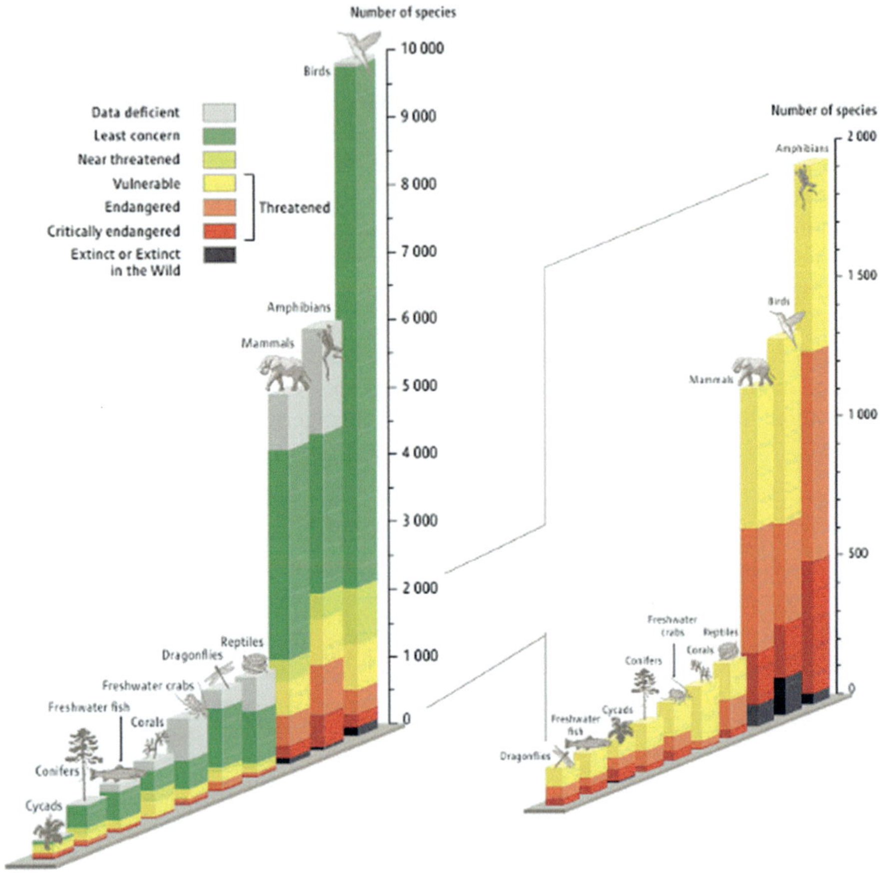

Fig. 7.2 Distribution of endangerment per type of species. (Source: IUCN)

Prioritisation issues follow a similar path. At a time where species go extinct faster than researchers can study them, which species should be studied first? Should we study the more useful species, whose protection is a priority to maintain the ecological equilibrium and biodiversity? It is a wise, but utilitarian stance. Should we instead prioritise the rare species, those that are the only representatives of their order, and are irreplaceable links in our understanding of living beings? Should we not start with the most endangered species, that need our scientific knowledge and conservation programs the most? Or should we instead focus on the most cognitively evolved ones, with complex emotions, long-term memory, self-awareness, an understanding of death, or, to put it bluntly, those that affect us the most because they resemble us?

The question of priority, although it does have very pragmatic origins and applications, raises an ancient issue that scientists were hoping to be done with: the value of beings. In the next point, we will discuss in more detail the question of complex species, and the added problems that come with them. But I believe it is an important part of this ethical examination to establish the fact that it does not necessarily make more sense to want to prioritise these species over other, further away from us, less anthropomorphic species but that are essential to biodiversity, such as bees or frogs.

And yet this is what we are tempted to do, and we think it is an interesting sign of the state of mind of research in general, as well as the perspective that we have more and more regarding complex species: the discoveries made about them have changed the value we give them.

This change in value results in new problems for scientists in this field, problems that did not exist a century ago and are new to the discipline. Problems about the distance of the researcher from their study subject, and their involvement. Once more, this line of reasoning came from human sciences, mostly social sciences but also language sciences (Grand Paris Sémiotique 2019): should the researcher be involved?

While objective distance was seen and accepted as proof of scientific rigor for centuries, the new situation, one of urgency and fragility of study subjects, transforms the link between researcher and subject. This transformation has not been very much discussed as of yet, although some works already exist (Bertrand and Darras 2019; Estay and Horrein 2019). I believe it is urgent to recognise that this issue exists, that it is a major one, and that the answers that we will be able to find (or not be able) will considerably influence the evolution of the disciplines involved in animal studies and the methodology in the decades to come.

7.2.2 About Complex Species

Studies on complex species have their own unique ethical issues. One of them, whether we are experimenting on a subject or dealing with the problem caused by the accelerating disappearance of specimens, comes under the concept of personality and the accompanying fact that, within these complex species, the individuals rarely are interchangeable.

Talking about animal personality was still inconceivable just a few years ago, even though the works of pioneers in primatology such as Dian Fossey (1970) and Jane Goodall (1986) already tended towards proving this was a particularly relevant term. However, while it is possible to study the semiotic manifestations of a species from a "social" perspective (by making a list of used signs, such as with an ethogram, by studying information transmission within the group and noticing a hierarchal system in which individuals play certain roles), it quickly finds itself having a limited reach.

This is all the more true in a work such as ours, as emotion is an individual experience, even if it does frequently spread to other members of the group, especially in very empathetic species (such as the gorilla and chimpanzee subjects of Fossey and Goodall). However, even when this type of situation occurs, individuals often do not react to the same emotion in the same way. De Waal described the final hours of Amos, the male chimpanzee, and explained that, while all the individuals in the group were aware that something very bad was happening, some were simply staring at Amos in silence while others were very agitated, and a few, such as a female named Daisy, were trying to help him by building a "nest" and prop him up using "pillows" made from wood wool (De Waal 2013, p. 43).

It is therefore primordial when working on complex species – particularly if they are endangered and only a few specimens remain – to not only take into consideration the species, so as to understand its general functioning, but to also consider the individual as such, that is to say a being that possesses their own personality, their own character traits, which often can only be explained by the subject "being that way". This choice is also coherent with the school of thought of biosemiotics that makes every effort to consider the subjects as the operators of their semiosis:

> But it seems obvious that evolution little by little created animals with central nervous systems to be players "in the ecological theater" (to borrow Hutchinson's famous phrase), the play itself changed character so that increasingly, evolutionary gains would turn upon the development of efficient mechanisms for social interaction and cooperation – as well upon such misinformative practices as cheating and faking and, in short, the evolutionary games would be expected to increasingly concern the acquiring of semiotic competence. (Hoffmeyer 2008 p. 188)

However, the current data in ethology going in this direction are still rather scarce (but some works such as those from Cédric Sueur (Sueur and Pelé 2016) showcase the differences, particularities and individualities as actually being intrinsic components of the species, and not secondary phenomena of no consequence.

There are, in my opinion, two very distinct reasons for this. The first, which is very understandable, is that ethologists do not pursue the same goals as zoosemioticians (who also are very scarce). They aim to understand the way a species operates, and while the special cases of certain individuals are enlightening or striking examples to them, they are not the main focus of their study, and therefore are rarely referenced as such. Thus, even with primatologists or elephant specialists that name the individuals they study, it has been difficult to find any "monitoring" of these individuals to study them as semiotic subjects evolving over time.

The second reason is what Frans de Waal calls the problem of the "human exception". This point of view, which has already been discussed in Chap. 2, becomes particularly harmful when studying complex, endangered species. While he first describes it as a phenomenon that is particularly rooted in human sciences, De Waal does not spare his own disciplinary field and mentions a telling example from his own work.

> *L'exception humaine demeure bien vivante dans les sciences humaines et les lettres. Ces disciplines persistent à résister aux comparaisons entre les humains et les autres animaux – même le mot « autres » les offusque.* (De Waal 2013 p. 143)
>
> *[...] pendant toute une décennie, on avait déclaré les humains uniques en leur genre parce que les chimpanzés, disait-on, ne s'entraidaient même pas quand cela ne leur coûtait rien. Nous avons amélioré la méthodologie, éliminé l'appareil, et voilà: nos chimpanzés s'entraident. Que des collègues viennent à présent objecter que toute l'affaire ne compte*

pas vraiment, c'est un peu fort. [...] Dès que le statut spécial de l'humanité est en jeu, les poteaux de but se déplacent comme par magie. (De Waal 2013 p. 170)

Translation:

The human exception still holds on in human sciences and the arts. Those disciplines persist in resisting comparisons between humans and other animals – even the word "other" offends them.

[…] during an entire decade, humans were considered an exception because, so we said, chimpanzees did not even help each other when there was no cost associated with it. We improved the methodology, removed the device, and there you go: chimpanzees help each other. For some colleagues to object that this does not really count is a bit much. […] As soon as the special status of humanity is put into question, the goalposts magically start moving.

For a long time, personality was not a serious research subject, or even a research subject at all, and therefore, was simply not researched. In the same book, De Waal explains that the discipline has been greatly correcting the course since the mid-2000s and that, as the old prejudices are being broken down, more and more interesting a complete studies in this field are appearing alongside a new generation of researchers.

Moreover, in a number of cases (Whiten and Boesch 2001), it seems that the disappearance of individuals, and then whole species, is not just a loss of biodiversity. Complex species can also take other things with them when they go extinct – techniques, knowledge, know-how, beliefs. When, in Sect. 3.2.4., the question of the relevance of defining the term "culture" came up, part of my reluctances (on top of the methodological ones, which have already been underlined) came from this situation, which is particularly problematic from an ethical standpoint: when these species are extinct, can we, should we, talk about the extinction of one or several cultures alongside them? Is culture preservation the responsibility of the researcher?

It should be noted here that I am not debating the question of captivity and its impact of this possible culture; while it is certain that a change in environment means a change in behaviour, but outside of deleterious captivity conditions leading to pathological behaviours, it is hard to say with certainty whether it is a good or a bad thing. De Waal made an interesting note on the subject in *De la reconciliation chez les primates*, explaining that captivity had the advantage of letting researchers observe individuals that were, paradoxically, "freer" because they were not trying to hide their emotional, cultural, technical etc. manifestations from possible predators, humans or otherwise (De Waal and Robert 2002).

7.2.3 About Pain in Animals

One last point remains regarding animal studies: the taking into consideration of pain.

This is of course a societal debate (laws against mistreatment have got harsher on a global scale in the last few years, and the amount of people becoming vegetarians or vegans has increased in the West) but also a lexicological one (in France, the debates on the terms of sensibility and sentience, even if the latter does not have a French equivalent as of yet, are very topical [Guillaume 2014]). This debate goes all the way to the industrial and agronomical circles, and is taking more and more room within the scientific sphere, on two different aspects.

The first aspect is science as a study tool of pain. Until recently, similarly to personality, pain in animals was very barely studied (a remnant of the "animal machine" school of thought), at least beyond its functional aspect. While nociception was a study subject, particularly in neurology and pharmacology, the impact of pain on animals, stress or non-physical mistreatment (shouting, threatening, scaring) were not really subjects.

This has changed nowadays, as researchers differentiate, for example, in zootechnics, animal welfare on one hand (not mistreating or depriving animals in farms) and well-being on the other (actively acting to provide optimal living conditions for the animal. The study of psychological or emotional pain, however, is still lagging behind.

During my thesis, I attempted to provide an example of method to remedy this lack of data by a semiotic study of trauma. This use of semiotic tools in their historical, original manner, which is to say an almost medical one, helps drawing conclusions of a nature close to the diagnostic.

Trauma is doubtlessly by far one of the strongest, most durable emotional manifestations. It is particularly useful for semioticians, because its intensity makes it almost impossible for an individual to be subjected to it and not show any external sign, and the fact that it is durable makes some of its signs perceptible in very specific conditions years after the event that caused it. Semiotics is a very adequate study tool for this type of emotional manifestation, as it can be very difficult without it to interpret a *seemingly* erratic behaviour or incomprehensible attitudes due to their temporal distance with the event.

My model, with the interpretation tools it possesses, provides interpretations that are plausible, coherent, and in the cases I submitted – thanks to the combined work of the other researchers – verifiable.

However, for these interpretations – and others afterwards on any and all emotional manifestations – to be considered valid, there can no longer be a lack of new perspective on emotion, more precisely on the biological role of emotions, and the potential psychological pain they can cause in the life of a subject. Especially since we know that, as each subject is a whole individual with their own behaviour and personality but especially their own emotions, each sign can change from one individual to another, as is the case with humans.

I observed that emotions, in the case of trauma, are not only very strong, but also what could be described as negative. This was not about value judgement, but rather about recognising that I was dealing with emotions caused by deprivation, resulting from the brutal disappearance of another individual: disappearance of affection, conform, tranquillity, that are now being replaced, after the split, by anguish, sorrow, a sort of fading away, the outbreak of a nervous disease or distress.

The strength of the trauma comes from the brutality with which the subject loses their emotional security, and shows the difficulty, if not impossibility, of living as long as this security has not been re-established. Even worse, this insecurity is so violent for the individual that it can affect them for the rest of their life, and influence their behaviour. This influence is not always negative – Fouts (& Mills, 1998, p. 238) attributed the very attentive care that Washoe showed towards her adopted son (attentive to the point that the young very quickly became able to master the signs his adoptive mother was teaching him) to the trauma she suffered from the loss of two very young children, one of which sent her into a very deep state of distress that worried the researcher. In this case, the trauma produced in the female, who had now recovered, positive caring behaviours. The fact remains that its influence was enough to push forward certain attitudes, years after it happened, and even after the female had finally been able to raise a child as she so dearly wanted.

It is no longer possible nowadays to look at such elements as secondary or not important when their influence is so large and they hinder survival as soon as they are not correctly satisfied. I believe that the use and efficacy of a semiotic model on emotion (for the part that involving us) speaks for itself when it is understood that emotion actually is a primary need in many species (especially the ones we discuss), without which individuals cannot properly grow and, sometimes, live. From then on, it is easy to understand why it is possible to still find interpretable signs months, even years after the facts that they are a reaction to, why it is so hard for individuals to not take their emotions into account (and therefore to not react and produce interpretable signs) and why emotional manifestations are omnipresent in the lives of certain species. All of these elements are arguments in favour of a larger, more involved study in psychological and emotional pain in non-human animals.

Studies on pain should therefore offer, just like in the field of human medicine, alternatives to pain.

Semiotically speaking, an element of answer appeared for me when, during a lecture at Lyon 1 University, in the veterinarian psychiatry class, I suggest to them not to consider the animal by itself, but to consider the animal and his or her owner as a hybrid semiotic being. A veterinary hospital is full of strong odours – of stress, urine, blood, alcohol – cries of pain or anxiety, but also full of stressed human beings. However, pets are selected by humans for their capacity to understand us, to empathise with us and in this situation, owners are often emitting stress semiosis. Veterinary students told me that, despise it is not advisable, they often practice vaccines, bandages or blood pressure test with the animal on the lap or in the arms of the owner, when they feel that he or she is placid enough to play an active role in the care. In this situation, the semiosis is still violent – animals are stung, their hurt limb is manipulated, they still smell fear and blood – but a counter semiosis is implemented that lowers the strength of the first one. By considering the animal and the owner as a whole semiotic entity, we understand that the owner semiosis is a part of the animal semiosis: if the owner is calm and confident, the animal, somehow, understands that despite the violent signs, there is no danger. Propositions to systematize this counter semiosis exist and I will approach them in the last section of this chapter.

On the opposite end of these studies on suffering is the second aspect of the debate, situations in which science is the source of pain.

Of course, we can think of experimentation animals, especially when it has a pharmaceutical aim, as it is currently subject to debates. Many safeguards were put into place in Europe to minimise this pain, and this can only be a source of celebration. First mentioned by Russell and Burch (1959), The Three Rs cannot be overlooked nowadays in laboratories. This rule tells researchers to systematically Reduce the number of animals involved in the experiments (no matter what they are (Festing 1998)), but also to try and Refine the experimentation so as to find the most appropriate methodology, which is also the least painful to the animals, and set limits that the experiment will not cross (Flecknell 1994). The Three Rs also require, as often as possible, to Replace the animal models with other solutions (in vitro, *in silico*, etc.). This is not always possible and this part runs into a number of issues that I do not have the ability to pass judgement on, considering the degree of specialisation the question requires. However, I can only applaud the will to research new alternative techniques to animal experimentation, such as, for example, organs-on-chips (see Fig. 7.3).

There is however an experimental field in which we can pass judgement on whether that pain is necessary: the behaviour studies field, which can be studied by semiotics. Despret (2009) writes that, while the experimentations of Harlow brought precious elements to ethology, is it really necessary to reproduce them on a regular basis, which would recreate each time the same traumas? Béata (2009, 2013) and Christen (2011) ask a similar question when they ask themselves whether there is no other possible method to prove that certain species – monkeys, in this instance – are altruistic than by using electric shocks (Wechkin et al. 1964). Could another protocol not have been found?

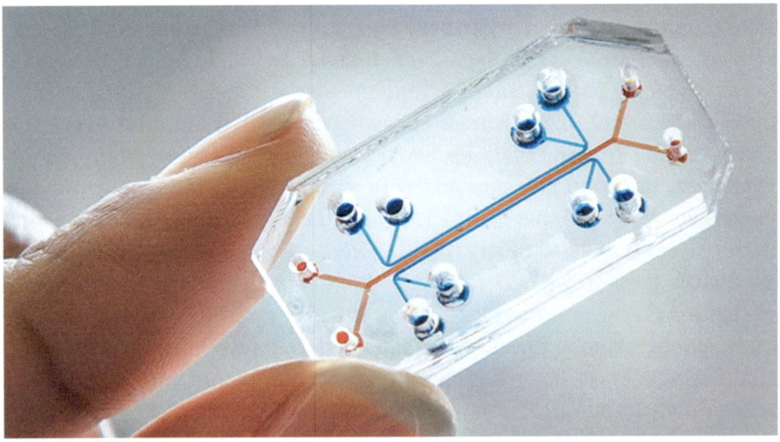

Fig. 7.3 Sample of human organs-on-chips

7.3 Solutions of Semiotic Methodology

In the last part of this chapter, we will now offer a view point and solutions from semiotics to the problems that came up in the first two parts.

First, we will focus on the general animal study problems that were discussed in the first part. Some of them (as well as the solutions that can be offered) may have already been mentioned in the previous chapters, but we will discuss them in detail here.

But what we will focus more on is the ethical aspects. Bioethics is a particularly fertile field when it comes to contact points between life sciences and human sciences. While philosophy and epistemology are the disciplines that are most often referred to, we think the inputs from semiotics are very real and appreciated, and we will be making them explicit.

7.3.1 On General Issues

Working with foreign minds, species that are not always close to ours, lead to a number of difficulties that we mentioned in the first part of this chapter which can have a semiotic solution. This entails focusing on the study of signs, of course, but also of the combinatorics of signs that carry meaning for the species involved; signs that, once connected, constitute the semiosphere of the species (Kull 1998), that is to say something that "*makes* sense" to that species.

To find something that "makes sense", semiotics is certainly better equipped than other branches of linguistics. For example, taking into account the complexity of a combinatorial is already an integral part of semiotics, especially in the Peircean school of thought. Although Peirce has already been mentioned quite a bit in this book, and his classification and sign component subdivisions demonstrated as more than useful, there is another thing to add.

The example of the works of Ryabov (2016) mentioned in Sect. 7.1.2 show that, for certain species, it is possible to talk about very elaborate forms of signs that are not all indicative of the same degree of complexity and do not all carry the same meaning. Precise work on the different categories of signs in Peircean semiotics and their characteristics when applied to ethology has already been done during our master's dissertation (Delahaye 2013).

This ability to take into account the various sign components will prove crucial in the future progress of zoology, semiotics and zoosemiotics (although we will discuss this in more detail in the next chapter), as it is very probable that more works like Ryabov's bring more and more complex sign combinatorics to light.

Beyond the question of language – which will not be discussed here any further than it has been in Chap. 3 – the question of Umwelt, of the mental world of the individual, of understanding and expressing them properly, is at the centre of the issue. The emotions, feelings, intentions have an influence on communication and the perceived meaning from the receiver. Take our species – it is absolutely possible to say a sentence while also making sure that our interlocutor understands without fail the exact opposite through irony or ambiguity: this is what linguists call illocutionary speech acts.

This part, which is not a linguistic production strictly speaking, but remains a semiotic production nonetheless, is primordial when it comes to understanding emotions and social ties. In his book *Je suis né un jour bleu*, writer Daniel Tammet (2009), suffering from Asperger's syndrome, explains how the internet saved his social life by clearly codifying the intentions and emotions of his interlocutors through the use of emoticons, while he had been unable to read those emotions until then.

Sign combinatorics is therefore a main focus when expressing and interpreting properly the content of the "black box" of the individual, to use the words of Eco (1999, p. 126). Nowadays, we see in ethology more and more cases where the combinatorics reaches a degree of complexity that can seem discouraging, which leads to a number of the problems mentioned in the first part of the chapter:

- While better attempts at taking into account priority senses should be commended, as said above, they do result in a new sort of confusion on the exact role of non-priority senses in the signification of the message (as a reminder, humans are also part of this problem, and neurosciences and psychology have long demonstrated the importance of physical posture, facial expression or tonal change in the process of understanding messages).
- A sense can be used for very different functions that should be sorted (this is the case in dolphins, who use ultrasounds to communication but also to orient themselves), which can be very complicated, as seen with the conclusions drawn by Ryabov.
- The same sense can emit (and receive) different signs that need to be classified and analysed differently (in dolphins, there is a difference between ultrasound pulses and the "clicks" that are in our hearing range, both of which are meant to be heard by their interlocutors).
- It is particularly complex to bring to light the mechanisms connecting different signs emitted by one sense. In the case of dolphins, are the "clicks" the punctuation of ultrasounds? Do they have, as our interjections do, a phatic value that aims to attract the attention of the individual we wish to speak to? Are they complementary signs that carry meaning, and if so, what falls within ultrasounds, what falls within "clicks", and why?

Here, my aim is to make us all think about the fact that the complexity of sign combinatorics may have been severely underestimated (including within our species). This thought process can only move towards a better cooperation between field disciplines and others, such as semiotics, that already possess complex analysis models that can, at least partly, answer this question. It is also the opportunity to remind that the life of beings is always full of signs, which do not all have the same importance, the same value and do not interact in the exact same way with each other. We should also put ourselves in the place of the studied individual and what carries meaning in this set of signs to them.

If at the beginning of this book, I insisted on discussing the notion of language (see Sect. 2.1.1) in the animal kingdom and separately discuss the notion of saying (see Sect. 3.2.1), it is because these are two distinct issues, in my opinion.

Language is a concept with a definition that varies from one scientific field to another, one school of thought to another; it can even change depending on the intentions of the researchers. Fouts (Fouts and Mills 1998, p. 105) wrote on the topic that the results reached by signing monkeys had led linguists to constantly redefine language so that it would remain exclusive to humans, until it was no longer possible to deny that some other species also have access to it, even in a hybrid form.

In my case, the choice was made at the beginning of this book to use a definition that seemed more objective and allowed to work unencumbered from the debates around the question that was the real focal point: how does an individual from a given species express to the other members of the species (and, in rarer instances, members of other species) what they have to "say"?

In a more general way, ethology and biology have made very interesting discoveries in the past few decades thanks to technological improvements, regarding communication through pheromones, postures, courtship dances or songs. But it still is difficult to determine which of those signs are using by individuals to "say".

On a less conscious (and the importance that unconscious signs can sometimes have has already been covered in this book) but similarly complex level, is the question of knowing what, amongst all the signs emitted by the other members of the species, is clearly a sign that makes sense for the individual. I do not need to perfectly understand what makes a painful emotion cause physiological disruptions that activate lacrimal glands to know that someone crying in front of me is distressed. Whether it is sadness, pain or rage, their tears "make a sign" and "make sense".

In my opinion, these are the basis of another proposed line of thinking regarding the challenges that come with studying animal semiotics, as this type of study can take many forms, involve many communication channels, complexity levels and degrees of consciousness.

To open up an avenue for a possible answer for these challenges, the goal was to show the possibility and usefulness of a bigger spot for semiotics within life sciences, in keeping with work like mine. This is done through two complementary aspects.

The first one is that it is becoming more and more crucial, as showcased by the works of Ryabov, to distinguish between different communication types of different importance within the studied species. The debate on language has seemingly damaged this, as it has led to classifying under the simple term of "communication" anything, according to this or the definition of language being currently used, that was not language without thinking of the relevance of such a classification.

Semiotics, mainly Peircean semiotics, can help with putting things back in order, not just on a form complexity perspective but also a complexity of conveyed meaning perspective. It is not possible to sort in the same category a sign showing sexual availability and a sign showing the emotion felt when touching the skull of what is recognised as a late beloved being. Far from the debate of what is language, why it is, and what is not, semiotics can be a precious ally with which to analyse the productions of studied individuals, give them a degree of complexity or cognitive demand, and then study them in the manner most appropriate to what these productions really are, rather than what the latest definitions in vogue would want them to be. Understanding foreign minds and different species also entails taking into account their subjective hierarchy of the world and how it is expressed.

The second aspect is that the study of signs can tell us a lot about the relationship that the studied individuals have with them, their expression and what they have to "say". Fouts credits the great success of Washoe learning ASL – and later on her song Loulis – to the importance that chimpanzees give to their exchanges with others, exchanges that need to be precise so as to translate their range of emotions, to clarify negotiations or to have others understand the rules of a game (Fouts & Mills, p. 238). The importance given to signs, the act of learning them, their transmission, their combinatorics when a specific sign is missing, are all semiotic study points that shine a brighter light on the mental world of subjects and the place that the signs hold in them, than they do on the type of signs being emitted.

In the case of my thesis, which was more focused on emotions, I could see that emotions are a major biological component of our species, as well as a growing but yet unknown number of other species. No longer can we pretend this aspect does not exist or has no impact on other biological components (reproduction, life expectancy, social life, etc.) simply because it is hard to measure, quantify and observe. Emotions are all the more important to take into account because their impact on sensitive cases (animals that are wounded, stressed, sick, in danger) is more and more recognised, and because cases in urgent situations, such as the ones in Sect. 7.1.3, are becoming more and more common. During my research work, I aimed to follow this path and say, in substance, "this thing exists, if it exists then it must be taken into account, and we should give ourselves the means to study it, which means we should build a model to do so".

It has not always been possible to create this type of interdisciplinary response focused on animals, and there have been several reminders in this book regarding how even today some data are still rare, scarce or just absent. But the advances of the various disciplines involved give more possibilities than even before. A particularly interesting aspect of these advances is that we are beginning to understand what we are missing to keep progressing. For example, Ryabov, when publishing his study on dolphin emissions, clearly explains the material limitations that cetologists are facing, and how necessary technological progress is necessary for this type of study to carry on and gather more reliable and numerous results.

Despite this, it should also be mentioned that work on conscience, memories or emotions can only be done with the full, constant knowledge of on major existing limitation: the enigma of individual experience. Because even amongst our own species and cultural group, it will always be impossible to completely understand the way another individual feels in the same way we understand how we feel. For example, emotions produce a large amount of manifestations that are diverse, rich in meaning, complex, and the choosing of semiotics to study this aspect of animal life is perfectly established by this point of view: the manifestations of emotions can tell us all of what is possible for us to know about said emotions.

What is "possible for us to know" is already very broad and largely unexplored. My belief is that there are many research opportunities waiting in this field. This is doubtlessly encouraged by a groundswell in research in the last few years that pushes for more and more interdisciplinary works. And as previously mentioned for my specialty, emotions are part of life, and life has difficulties fitting into fixed categories of taxonomy, lexicon or academia. Work mixing different disciplines are the best equipped (and possibly the only ones equipped) to efficiently study this type of phenomena, simply because these are tightly intertwined with other aspects of life (cognition, morals, memory, social ties, aesthetic sensibilities, reproduction, etc.).

The "final" limitation of work on animals should not distract us from all the work left to accomplish to understand what is possible to understand regardless. The challenge of knowing these "possible" things should not make us think they are not necessary and important.

7.3.2 *On Ethical Issues*

Regarding the ethical aspect of the issues that have been discussed, I believe a number of them can find satisfactory, although non-optimal, solutions by taking into account the semiotic dimension of the animal and their study.

We had previously discussed the case of the researcher facing the issue of species being very fragile nowadays, whether they are species being threatened in the wild, individuals being treated in reservations or populations in various reproduction programs.

In all these possible configurations, humans still have common parts to play. The first one is the protection of the individuals, often victims of poaching, pollution and deforestation. There is no real problem with this point, as long as we remember that animals, once again, can be unaware of which human being is an ally and which one is a threat. In the wild, this implies reducing contact with animals as much as possible, to avoid making them too familiar with humans, and then easy targets for poachers. In captivity, we must remember that some of these animals are traumatized by previous human interactions – no matter if they were benevolent violence or real violence – and cannot make the difference between these previous humans and us, potentially leading to aggression.

It is for the cases most tightly linked to captivity, especially reproduction programs, that we have started to set up the concept of "counter-semiosis", an attitude from the researcher or benevolent human involved with the animal who is trying to counteract the deleterious effects of natural semiosis of the human-animal relationship, where the latter often sees the actions of the former – which they cannot understand – as a threat. This is the case, in reproduction programs, of organised separations of individuals to avoid consanguinity but can sometimes lead individuals into great emotional distress.

Of course, this distress can become totally counterproductive: it is damaging the social fabric of the group and can cause the reproduction rate to drop. In this situation, it is not really possible to create a satisfying counter semiosis, the loss of an individual is a loss, no matter what we do. But taking into account the individual Umwelt forces us to look differently at the collective picture: is this precise situation, between these two precise individuals, is the inbreeding risk superior to the severe distress risk. We must acknowledge that what we are doing to them is violent; benevolent violence, but violence nonetheless; only after that can we honestly decide if it worth it or not to make the best decision possible.

Our benevolent interactions with animals need to take more into account their Umwelt precisely because not doing so can lead our interaction to fail. It is not enough to say that we are doing so for the animal benefit and that the end justifies the means, precisely because the means can compromise the end if we are not careful enough. By engaging a potentially violent interaction with a sensitive being, we must try to understand how this semiosis makes sense in its Umwelt and how we can emit a counter semiosis to reduce the violence's signals.

This can be done in a minimal way, a "lesser evil" choice. In the case of reproduction programs, taking the Umwelt into account should lead us to make different decisions, to decide less easily to displace a breeder and to consider the point of view of individuals as much as collective necessity.

In the cases when this violence is necessary, simply knowing that it is violence should lead to special behaviour from humans to lower the distress and stress level of the individual. For the ones in captivity, having more contact with caretakers, receiving their favourite food and being non-exposed to public – in the case of zoos – for a "healing" period can be signs of comfort and help a necessary transition.

This counter semiosis can also be much more present, for animals already living close to humans. For example, some veterinary hospitals are trying today to implement new features helping to reassure hospitalized pets. It could be special rooms, dedicated to visits, where the owner and his or her animal can spend time, grooming or playing. During these visits, owners are encouraged to bring favourite toys or blankets to be kept in the cage with the pet. Introducing the owner and toys in the hospital is a way to counter a large anxiety signal: the clear difference between the hospital – with unknown humans, unknown smells and pain – and the house – with the owner, familiar objects and security.

By countering signals, we are in fact countering the meaning these signs have in the animal Umwelt, the meaning of danger and insecurity. This counter semiosis can also be done through unnecessary interactions that only bring comfort to the pet – giving warming blankets, grooming, playing softly with the healthier ones, talking to the ones in intensive care. If all of this can seem superfluous, we must not forget that they are, in fact, playing an important part in reducing stress and pain, and therefore, in raising the survival rate or decreasing the aftermath.

That is why, as our benevolent but invasive interactions with animals are rising, we must more than ever understand that taking care of the physical animal is not enough to take care of the animal in a whole. Some of our interactions are violent but they are not less necessary, they must be done, and because they must be done, we must find a way to do them better. Taking into account the animal Umwelt is starting an honest reflexion on what semiosis we are emitting during our interactions, how it makes sense for the animal we want to interact with and how we can work on this perceived sense by working on our semiosis. If we want to take care of animals, we must understand that their semiotic world is also an aspect to consider.

To conclude this chapter, I would like to discuss the issue of science creating pain in animals that was previous brought up. I do not have a concrete solution to this problem, as I believe it is mostly a question of reassessing moral systems and values scales. Is it more important to develop a new medication or to not have any test subject suffer? This question is laden with implications and falls under social debate rather than the work of the researcher. Which is not to say the researcher does not have any impact on the topic. But it seems to me that, with the current state of knowledge and technical means, The Three Rs already provide an important point of reference to researchers that tells them not to needlessly cause pain in their work, especially if that work is with complex species.

References

Association Grand Paris Sémiotique. (2019). L'Engagement. *Confrontations sémiotiques*. To be published.

Béata, C. (Ed.). (2009). *Zoopsychiatrie: L'attachement*. Marseille: Solal Editeurs.

Béata, C. (2013). *Au risque d'aimer*. Paris: Odile Jacob.

Bertrand, D., & Darras, B. (2019). In Association Grand Paris Sémiotique, *L'Engagement. Confrontations sémiotiques*. To be published.

Christen, Y. (2011). *L'Animal est-il une personne ?* Barcelone: Flammarion.

De Waal, F. (2013). *Le Bonobo, Dieu et nous*. Lonrai: Les Liens qui Libèrent.

De Waal, F., & Robert, M. (2002). *De la réconciliation chez les primates*. Paris: Flammarion.

Delahaye, P. (2013). *Application de la zoosémiotique à l'éthologie: peut-on parler de culture animale ?* Paris: Université Paris V René Descartes.

Despret, V. (2009). Il est temps d'en finir avec Harlow. In C. Béata (Ed.), *Zoopsychiatrie: L'attachement* (p. 13). Marseille: Solal Editeurs.

Eco, U. (1999). *Kant et l'ornithorynque*. Paris: Grasset.

Estay, V., & Horrein, R. (2019). Reculer pour mieux s'engager ? Sémiotique et Cultural Studies. In Association Grand Paris Sémiotique, *L'Engagement. Confrontations sémiotiques*. To be published.

Festing, M. F. W. (1998). Reducing the use of laboratory animals in biomedical research: Problems and possible solutions. *Alternative to Laboratory Animals, 26*, 283–301.

Feynman, R. P. (2000). *Le cours de physique de Feynman 5: Mécanique quantique*. Paris: Dunod.

Flecknell, P. A. (1994). Refinement of animal use – assessment and alleviation of pain and distress. *Laboratory Animals, 28*, 222–231.

Fossey, D. (1970). Making friends with mountain gorillas. *National Geographic, 137*, 48–67.
Fouts, R., & Mills, S. T. (1998). *L'école des chimpanzés: ce que les chimpanzés nous apprennent sur l'humanité*. Paris: J. C. Lattès.
Goodall, J. (1986). *The Chimpanzees of Gombe: patterns of behavior*. Boston: Bellknap Press of theHarvard University Press.
Guillaume, A. (2014). Animal: « être sensible » unanimement désensibilisé. Sémiotique du sensible. *Revue trimestrielle de la Fondation Droit Animal, Éthique et. Sciences, 81*, 35–37.
Hoffmeyer, J. (2008). *Biosemiotics. An examination into the signs of life and the life of signs*. Scranton: University of Scranton Press.
Kull, K. (1998). On Semiosis, Umwelt, and Semiosphere. *Semiotica, 120*(3/4), 299–310.
Russell, W. M. S., & Burch, R. L. (1959). *The principles of humane experimental technique*. London: Methuen.
Ryabov, V. A. (2016). The study of acoustic signals and the supposed spoken language of the dolphin. *St. Petersburg Polytechnical University Journal: Physics and Mathematics, 2*(3), 231–239.
Smith, A. (Yankelovich, D.). (1972). *Supermoney*. New York: Popular Library.
Sueur, C., & Pelé, M. (2016). Social network and decision-making in primates: A report on Franco-Japanese research collaborations. *Primates, 57*, 327–332.
Tammet, D. (2009). *Je suis né un jour bleu* (p. 84). Paris: Éditions.
Wechkin, S., Masserman, J. H., & Terris, W. (1964). Shock to a conspecific as an aversive stimulus. *Psychonomic Science, 1*, 47–48.
Whiten, A., & Boesch, C. (2001). The cultures of chimpanzees. *Scientific American, 284*, 60–67.

Chapter 8
Building Zoosemiotics

Abstract Evaluation sub-question n°2: finally, what are the stakes of a real zoosemiotics discipline, and why build it?

This last chapter aims to explain how the stakes of zoosemiotics operate, by presenting the links of stakes that unite semiotics with biosemiotics, and the latter with zoosemiotics. It also gives an overview of progress in the field: progress that was made – where we stand now – as well as upcoming progress – which way it is heading – and progress that we hope for – in what direction it should evolve. The last part of this chapter is meant to constitute a conclusion to this book and to summarize its statements.

Keywords Biosemiotics · Zoosemiotics · Semiotics methodology · Interdisciplinary methodology · Interdisciplinarity · Animal studies

In this last, closing chapter of this methodological book, the aim is to deal with the second evaluation sub-question as defined in Chap. 1: finally, what are the stakes of a real zoosemiotics discipline, and why build it?

Of course, zoosemiotics is a term, as well as a discipline, that has existed since Sebeok (1968, 1973); the act of building it is not a completely new creation, so to speak. But is clear that there have been few continuators in this specific field, and zoosemiotics is still, as a result, a young discipline looking for its framework and landmarks.

In the first part of this chapter, I will showcase the disciplinary developments that exist or tend to exist nowadays between semiotics and animal studies, what they bring, and what their challenges are.

This will be done by first discussing the particularly strong, refined bridge that exists between semiotics and biosemiotics (Hoffmeyer 2008; Favareau et al. 2012). Then we will move on to the rarer, but already interesting and worthy of attention, bridges between biosemiotics and zoosemiotics.

The second part will focus more on the progress that still needs to be made, the paths that which zoosemiotics wants to explore and the aspects it should set its sights upon.

P. Delahaye, *A Semiotic Methodology for Animal Studies*, Biosemiotics 19,
https://doi.org/10.1007/978-3-030-28813-6_8

This will be done by first making clear where we are in this discipline of zoosemiotics, to paraphrase the identical work of Kull (2012) on biosemiotics. Then I will attempt to determine, according to what was done in Chaps. 2, 5 and 6, where zoosemiotics is currently heading and what that says and about its future as a discipline. This will be treated as a separate point from the next one, which is, what is it expected to do, and where should it go – this reasoning will be based on the issues discussed in Chap. 7.

Finally, the last chapter will be used to provide the conclusion of my work and this book, as I deemed a separate chapter for this to be superfluous, choosing instead to put the conclusion of this book at the end of an open-ended chapter resolutely pointing towards the future of the discipline.

8.1 Between Semiotics and Animal Studies

In the first part of this chapter, we will study the shared challenges between semiotics – I could even say the different types of semiotics – and animal studies.

This part is an opportunity to detail the reasons and challenges that lead to not only create animal study models from the methodology in this book, but also use them in broader and broader fields and to answer more and more varied question, in diverse academic fields.

Firstly, I will introduce in a more specific manner the challenges that exist in the connection between general semiotics and biosemiotics. With the latter obviously is a branch of semiotics, and uses its tools and concepts, the study subjects it focuses on, which often are very different from the majorly textual tradition of semiotic analysis (especially Saussurian or Greimassian semiotics), make it a separate discipline with its own objectives, methods and challenges.

Secondly, I will present, in a similar manner to the first part, but using an opposite starting point, the challenges that connect general biosemiotics to zoosemiotics. Both disciplines directly fall under semiotics and share a great number of common characteristics. Their connection can be interpreted differently whether you consider zoosemiotics as a branch of semiotics, on a similar level as biosemiotics, or you consider it as a branch of biosemiotics that focuses on a specific aspect ("zoo") from a general subject ("bio").

8.1.1 Semiotics and Biosemiotics

Speaking of semiotics as if it is just one unit sounds like an abuse of terms. It would be far more accurate to speak of types of semiotics, to not only properly take into account the various branches, but also the various models and theoretical schools.

It is difficult to talk about all semiotic theories, as some are, according to certain semioticians, incompatible in their approach (Bertrand and Darras 2019). On the other hand, it should be wondered what schools of semiotics have contributed, or could contribute, to biosemiotics.

While I have already discussed in length the topics of Peircean semiotics, pragmatism and what all of their tools and concepts allow us to do in biosemiotics, I have yet to deal with the question of the shared challenges between this branch of semiotics and the biosemiotics branch. The Peircean school of thought, in semiotics, has a broad aim: understanding the different forms of semiosis, regardless of their production origin, as opposed to Saussurian or Greimassian semiotics, which mostly focus on textual productions and articulated language, although not exclusively.

This was all biosemiotics needed to become a living, developing discipline and see information exchange systems in all living systems (Hoffmeyer 2008). These information exchange systems can be more or less complex, more or less permissive, more or less creative, more or less interconnected. It goes without saying that the aim is not to study the semiosis of the transmission of pharmacopoeia in great apes[1] in the same way as the semiotic interaction between cancerous cells and healthy cells.[2] But Peircean semiotics is doubtlessly the one that equips biosemiotics the best when it aims to study and analyse the different semiotic degrees of the living. For some researchers,[3] the Peircean school of thought is even the only one that allows for such semiotic adaptability.

Does this mean that there are no inputs, no aspects from other branches of semiotics that can add to the thought that goes into the methodology of biosemiotics? Certainly not – but we are dealing with disciplines that have different fields of action.

In chapter, the question of involvement, of the potential implication of the semiotician with their study subject was raised for researchers in biosemiotics or zoosemiotics, but it should be mentioned they are not the only ones for which the question may be asked. In more general semioticians, this question is in full swing, and the colloquiums around this theme[4] gather researchers from every school of thought.

When confronted with subjects that are complex, yet classical in their form, (a textual or iconographical study for example,) with powerful content (war memories, endangered languages, etc.), Saussurian or Greimassian semioticians end up dealing with issues close to those discussed in Sect. 7.1.3., with the current impossibility

[1] P. Delahaye "Zoosemiotics of emotions: A new model and its applications", 18th Gatherings in Biosemiotics, University of Berkeley, California, 2018.

[2] J. Buttlar "Metastasis of pancreatic ductal adenocarcinoma – convention and fraternization between cancer cells and others", 18th Gatherings in Biosemiotics, University of Berkeley, California, 2018.

[3] S. Brier "It is Peirce all the way down: Why biosemiotics needs a Peircean triadic semiotic pragmaticist process philosophy to be transdisciplinary scientific", 18th Gatherings in Biosemiotics, Univeristy of Berkeley, California, 2018.

[4] A recent example is from the *Grand Paris Sémiotique*, an association of several Parisian universities around the different types of semiotics, which I have the honour of being one of the members of the board of directors: "Sémiotiques: implication et engagement". Association Grand Paris Sémiotique (2019).

(or at least great difficulty) of maintaining the traditional objective distance asked of the scientist from their study subject.

Therefore, while the aspects that are specific to biosemiotics can only really take from the tools and methodology of its Peircean branch, we should not neglect the aspects that are specific to the semiotic discipline, for which the interaction between schools of thought can be, and must be, strong if we want general solutions to emerge.

It goes without saying that, regarding aspects specific to biosemiotics, tools and methods can also come from other disciplinary fields, mostly life sciences, but this has already been discussed in Chap. 6.

8.1.2 Biosemiotics and Zoosemiotics

Works in zoosemiotics and biosemiotics (the relationship between the two will be discussed later) are the best bridge between language sciences and life sciences. Thanks to this first effort to have the two disciplines coexist, the founders of zoosemiotics have already reconsidered the concepts of sign and semiosis as they were already majorly defined within general semiotics, which is far from negligible in a work such as this:

> In asking whether semiotics is coextensive with biology, we have already challenged dominant glottocentric notions of semiotics. It is one thing to challenge, however, and quite another to vanquish. (Deely 1992 p. 45)
>
> What then is the sign? It is simply the element that is playing the role of a standing-for at any given time. (Deely 1992 p. 47)

While there necessarily are many shared concepts and tools between biosemiotics and zoosemiotics, there exist enough differences between the two disciplines to ask the following question: should biosemiotics and zoosemiotics be considered as two sister branches that both come from semiotics, or should zoosemiotics be considered as a sub-discipline of biosemiotics, an academic domain tasked with studying a specific point of its mother-discipline?

This question may seem anecdotic, but since we are discussing the building of a relatively new discipline, it is necessary to know the heritage and foundations we are building on as accurately as possible.

The answer that comes naturally is that zoosemiotics should be considered as a daughter discipline to biosemiotics. This is perfectly logical if we look at the life sciences component of both disciplines, as the "zoo" is part of the "bio", a more general term that encompasses other categories of beings, other forms of possible semiotics.

From this point of view, it is not unreasonable to think that, while it has the goals of its mother discipline – that is to say, understand, model and analyse semiotics in living beings in all their wealth of diversity – zoosemiotics has been made to go it alone, and create its own tools and methodologies due to the particular complexity and broad diversity that exist in the animal kingdom (including humans or not, depending on the perspective).

Yet I am not certain I share this analysis. While it is clear that the zoosphere is only a part of the biosphere, the current definition of the relation between zoosemiotics and biosemiotics brings problems on the levels of methodology and the history of concepts.

A daughter discipline inherits goals, concepts and methods from its mother discipline, which it then changes according to its own specificity, but they still exist as a conceptual starting point. However, in the case of these two disciplines, it appears that the story went the other way around.

While the earliest occurrence or the term "biosemiotics" we could find goes back to 1962 (Rothschild 2000), the concept that truly gives birth to zoosemiotics, the concept of Umwelt, appears as early as 1909 (von Uexküll 1934). It is therefore problematic to say that a discipline that has *seemingly* appeared in second is the mother discipline of the first. More problematic even is the fact that it cannot be said that there was an inheritance of concepts and methodologies, as the main concepts and tools of zoosemiotics are technically older than the first appearance of biosemiotics as a discipline.

As a result, in this instance, it would be more sensible to consider the two disciplines as branches of general semiotics on the same level, with their own growth and issues.

I would like to provide a third path of analysis which I believe to be more fertile for the current building of zoosemiotics.

The works of von Uexküll (1934), just as the works of Sebeok (1973) would be later on, are not really specific to zoosemiotics. They can be seen as more general, with the goal of extracting from general semiotics a branch that would analyse and interpret the different semiotic manifestations from non-human living beings, as an attempt to open up semiotics to other kingdoms. However, as the animal kingdom is the closest to us (since we are an integral part of it) and the one with the most obvious semiotic manifestations, the zoosemiotics part would take over, while biosemiotics, with the rest of living things, would remain in the shadows for a while longer. Let's not forget (Chap. 1) that up until recently, the study of animal semiosis was seen as a far-fetched endeavour that could not be further away from science. So what would be said about the semiosis of other kingdoms?

Therefore, here is my analysis: the main tools and concepts of biosemiotics and zoosemiotics were probably created at around the same time, even if the former was not mentioning its name just yet. It is therefore normal to think that both disciplines have similar goals and issues, because they have a strong, shared heritage. However, as zoosemiotics had to go it alone for a certain amount of time, it developed its own methodologies and more specific goals and issues relating to its subject matter. So it cannot be considered a daughter discipline of biosemiotics, although it is logical, from a phylogenetic perspective, we could say, to see it as a sub-discipline of biosemiotics since animals are only a part of living beings.

Zoosemiotics is therefore tightly intertwined with biosemiotics, it shares its challenges and concepts, and even though it is a sub-branch of the latter, it still is an autonomous one, and any difference in methodology or conceptual tool is usually justified by its history, and should not startle researchers going back and forth between the two disciplines, or between life sciences and one of the two branches.

Some harmonisation (of definitions, concepts, tools) will sooner or later be desired here as well.

8.2 Progress Wanted, and Progress Needed

The second part of this chapter will focus on the current state of zoosemiotics and the reasons why it should continue to grow.

Logically, this will be done by first drawing up a report on the state of zoosemiotics (although its state in not permanent, as it is a relatively new discipline, and therefore evolves and grows very quickly): what has it allowed, what has it created, where are we with this discipline and its study subjects?

Based on this report, I will provide a projection for the years to come. How does the discipline tend to evolve, what are we going towards, what can be said about its focal points, uses and the progress expected from it in the future?

But is where we are going where we *should* be going? At the end of the chapter, I will discuss the use of the discipline and the coherence between the challenged mentioned in the previous part, and the path that zoosemiotics is currently on. How do you build a discipline that is coherent with its challenges, methodologies and study subjects? What advances do we want to see in zoosemiotics, and which ones do we need to make?

8.2.1 Where We Are

Zoosemiotics is currently a discipline that grows very quickly and is garnering more and more interest. Theses on the topic have been multiplying over the past few years, whether they directly mention the discipline (Delahaye 2017) or not (Rattasepp 2018), or sometimes are in an intermediate area (Mäekivi 2018).

Yet it is still difficult to place it in the academic space: works that claim to be part of it often come from research departments in linguistics or communication, and sometimes in semiotics, but such departments are still rare, especially in France.

The works themselves are not also easy to categorise, as they are sometimes considered a part of human sciences, or sometimes ethology. Their deeply interdisciplinary nature is rarely taken into account. Due to a lack of substantial training regarding interdisciplinarity and intertheoricity, this nature is misunderstood and badly integrated by the more traditional academic institutions.

There is still quite a bit of reluctance within human sciences regarding these works, even if it has been established already that this is due to historical and ideological reasons most of the time, rather than a true methodological calling into question. However, this type of behaviour tends to be lessened in humanities which are already in contact with animal issues, such as philosophy (in general, but also more specifically certain branches such as bioethics, philosophy of law, etc.).

A similar attitude seems to quickly evolve within life sciences which, while they might have violently rejected zoosemiotics these in the past (especially during Skinner's time or during the behavioural hegemony (Machamer 2009)), seem more and more partial to the new tools, concepts and methodologies that zoosemiotics can bring.

Scientific life also appears to develop around the discipline, or at least around its issues. There are no (at least not yet) zoosemiotics colloquiums, the studies of this discipline have been welcomed with open arms for the last few years in the Gatherings in Biosemiotics (Rattasepp and Bennett 2012), and other colloquiums that offer interdisciplinary perspectives to semiotic studies are beginning to appear (Delahaye 2018a).

A number of initiatives, seminars, partnerships and projects have been built in the last few years with a zoosemiotic component, and these initiatives, mostly broadcast in France by the very young *Société Française de Zoosémiotique* (Delahaye 2018b), tend to be integrated more and more within the landscape of research. This type of project, which deal with issues that are of great concern to general society as of late, help creating a connection between the general public and the world of research, which is often seen as very inaccessible.

A report on the current state of zoosemiotics can be done by describing three different aspects.

The first one is that it is a relatively young discipline which answers very current issues, and as a result is a very dynamic and fertile discipline with a lot of innovation and developments in many different directions. While there is a risk of spreading the discipline too thin, it is clearly not the case for now, and the relatively low number of researchers in the discipline should ensure a certain coherence in the issues being studied.

The second aspect is a knock-on effect of the youthfulness of the discipline. Due to it being relatively new, it is still badly represented in academic institutions, lacks recognition, visibility amongst peers as well as funding, despite academic policies advocating interdisciplinarity more and more.

The last aspect is that the discipline boasts a great amount of interest from the general public and civil society due to the issues it deals with. As it is particularly well-equipped to create a bridge between them and the academic field, it is more and more represented, and invited, in science popularisation events and initiatives, which makes it more well-known and popular in general, but consequently, also more visible to the rest of the academic world.

8.2.2 Where We Are Going

While the report on the current state of the discipline may seem mixed, but mostly positive, it does not completely show the direction in which things will evolve regarding the growth of zoosemiotics.

The discipline is clearly at a point in its development where its advances are full of promise and suggest positive developments, but its youth and a certain academic fragility could still clip its wings. As a reminder, after the last publications by Sebeok using the term (Sebeok 2001), the discipline appeared to have disappeared. It is still possible nowadays for something similarly unfortunate to happen again.

However, zoosemiotics is not currently on this path, and instead, it can be said that it is heading toward a growth of the discipline. While still being wary of over-confidence on the subject, this hypothesis can be chosen for three main reasons (others are possible, but I believe that these three carry the most weight in the evolution of the discipline).

The first reason is that this time, the discipline writes, studies and proposes answers in a better synchronised manner. Animals, the way they feel, their development, their well-being, their emotions and their communication have become true scientific study subjects in fundamental research (especially on the development of cognitive abilities), research applied to veterinary sciences or agriculture (with the development of studies on animal behaviour and well-being), and even in research in human and social sciences (mostly in the fields of philosophy and ethics, but also in law or psychology). With increasingly important and specific demands on animals and their functioning, zoosemiotics will surely find a stable, perennial place more easily than it could have hoped to at the time of the first publications of Sebeok (1973) on the subject.

The second reason is that the discipline is garnering more and more legitimacy with civil society. While it has traditionally been indifferent, if not hostile, at the idea of being compared and approximated to animals due to its religious heritage (at least, in the West), civil society has drastically changed its opinion in the last few years, demanding more well-being, rights, and juridical protection for animals. This approximation has also led to therapeutic innovations, such as clinical trials in zootherapy, which would not have been conceivable a few decades ago. This societal enthusiasm, which has given more and more importance to animals, demands studies and specialists to fulfil these expectations. As a result, the development of zoosemiotics arrives at a time where the interest for its subject matter and the work it can provide is particularly appreciable.

The third reason to finish this development prediction is the one that has been detailed in Sects. 7.1.3 and 7.2.1, the current emergency situation in which biodiversity is. Animal studies not only become more and more difficult as the population of some subjects lowers and as the ecosystems become more and more affected, they also receive very substantial goals. Understanding the way species function to preserve them, to integrate them into reproduction programs, to re-introduce them into the wild, is becoming massively important, especially since the window in which to act seems to be getting shorter and shorter every year. This may be a cynical observation, but animal studies will need to work with every existence forces if they wish to measure up to the challenges they face, and zoosemiotics is developing in that wish for a general gathering of disciplines.

It is for all these reasons that I believe the development of the discipline will be going upwards as it meets the needs of current preoccupations on its research subjects; but this development has a lot to do with timing.

8.2.3 *Where We Need to Go*

While all of these facts are a clear indication of the direction in which the discipline is likely to evolve, and on what bases, they do not tell us the way in which it *needs* to go.

As underlined in the previous chapter, animal studies are now a field that tends to involve the researcher more and more. In this type of situation, with large challenges and circumstances that sometimes are emergency situations, it is not unreasonable to wonder the way a discipline *should* evolve, even if the question can seem iconoclastic.

So, what meaning should be given to the semiotic study of animals? Firstly, while it must be said that ethology, amongst other disciplines, has made numerous and particularly interesting discoveries on animal in the last few years, it also has, as every discipline does, its limits. In a seminar for the *Société Française de Zoosémiotique*, Michel Kreutzer,[5] an ethologist and ornithologist, explained that ethology has nowadays understood and identified its biggest limits, and that other disciplines should take the reins once it is reached. One of these limits is understanding precisely the process in the brain during a certain behaviour, which now falls under the umbrella of neurosciences, another is understanding the impact and previous history of a behaviour from an animal. This limit does not officially have a discipline that can take it over but zoosemiotics, through the very nature of semiotics (studying signs to find meaning) seems very well equipped to deal with this.

This is all the more true as a semiotic study on animal, as demonstrated throughout this book, must take into account multiple disciplines and academic fields, harmonise its own knowledge with the pre-existing definitions and methodologies from other specialties dealing with animals and innovate in the creation of truly interdisciplinary models and theories. Zoosemiotics has nothing to gain from evolving separately and from studying animals using models that are its own and different from the others in the field on purpose. On the contrary, my belief is that it must strive to fill that empty space drawn by the limit recognised by ethology to gain new elements from it, use it as a support and provide it with new elements in return.

[5] M. Kreutzer « Histoire de l'éthologie », Séminaire de la Société Française de Zoosémiotique, Sorbonne Université Paris, February 2019.

If I believe this is the direction in which zoosemiotics *needs* to go, it is not just because it is the most obvious, "ready-made" one but rather because it is the place in which the discipline will be at its most useful. It has not always been easy to have people agree that animals should be studied for themselves, and not just as a comparison point to humans. Now that this idea has become more present, I believe it is important to put zoosemiotics in the great academic field of animal study, with this strong driving principle: semiosis is a major component in the life of many species of animals, and it is no longer possible today to try and study them while leaving out this aspect.

And I believe that this place of a study discipline on a major aspect of animal life must be asserted very strongly due to the fact that many species, especially amongst the most complex or those with the most specific behaviours, are currently severely endangered. To not be able to or not want to study them now leads to a risk of seeing this knowledge disappear along the last members of these species.

8.3 Conclusion

I hope that in this book, I was able to show that a semiotic methodology of animal studies can provide the three types of contributions explained in the introduction (Chap. 1): linguistic, ethological and societal.

Language sciences have everything to gain by getting rid of past reluctances to focus on the study of non-human animals, to acquire new knowledge as well as new methodological perspectives and, I hope, a better representation of semiotics in these disciplines. Life sciences, on the other hand, with this type of methodology, could see the arrival of tools and conceptual models they did not possess and that they were surely lacking to explore certain aspects of animal life intertwined with semiosis. Regarding the societal contribution, this chapter already discusses the topic quite a bit, so I think there is no need to return to it.

I hope to have shown in this book that the conception of a methodology of this type is not done carelessly, and that it entails taking into account a number of elements that seem, at first glance, to not be essential to scientific work but could end up interfering with it if not studied meticulously.

The debates and controversies, while they may seem more like ideological quarrels rather than scientific ones, have forged the history of the discipline and act as a backdrop – that we are not always aware of – to the creation – conscious or not – of any new methodology that offers to study various animal species through the use of tools from life sciences. To not address these debates, even simply to make it clear that we are distancing ourselves from them and why, poses the risk of seeing them resurface in the midst of work. I believe it is wiser to acknowledge them and take their existence into account in the initial stages of any semiotic study of this type.

As many of these controversies crystallise on definition issues (Chap. 3), a semiotic methodology for animal studies should begin with a meticulous endeavour to harmonise them from an interdisciplinary standpoint, to also avoid the issues connected to different acceptations of the same term by several of the disciplines involved. It should be studied why certain definition issues cannot be solved in the current state of things, or why certain definitions carry meaning that is too ideologically charged. It should be specified, at the beginning of the semiotic work, why certain terms will not be used, which terms were chosen over them or, on the contrary, why certain terms will be used with this specific acceptation of the meaning, and why it does not include certain specific semes.

Lastly, as the subject is a semiotic work applied to life sciences, the choice of the tools and concepts that will be used (Chap. 4) is primordial. The ones I chose to present not only seem to be the better adapted but also the more polyvalent ones. The work I have already done using them leads me to think they are able to provide real, strong methodological support in a variety of species, situations of observation or experimentation, and precise study subjects (communication, emotion, social life, etc.). This methodological model remains open-ended, however, and it is perfectly possible that, in certain circumstances, for certain cases, other semiotic tools and concepts need to be included. In this case, I hope to have provided, especially through the use of relevance theory (Sperber and Wilson 1989), ways to choose and sort these tools in the best possible manner so that they can be relevant to the studied animal.

I also hope to have transmitted throughout this book the different steps that allow the creation, the building of a methodological model that allows for semiotic analysis of animal productions. These are the same steps that have always ruled over the creation of methodological models, but the necessary interdisciplinary aspect of this type of work makes them broader, more complex, with a larger number of parameters that need to be taken into account.

In this way, the presentation of the concept of intertheoricity (Guillaume 2013, 2014) as a way of building bigger theoretical models (Chap. 5), is a can't-miss aspect of these creation steps. It of course entails that the definitions between the disciplines involved have been previously harmoniser, and that the right tools and concepts that are the most suitable to be exported and developed within another academic field, thanks to their plasticity, have already been chosen. This step of creation and hybridisation of theoretical frameworks to get to a new, broader methodology, cannot dismiss the steps described in the previous chapters.

Similarly, the creation of a new methodological model entails knowing where to place it amongst what already exists within the field, in our case, the field of life sciences. To propose a semiotic study of animals is knowing precisely the strengths of life sciences, with which it is important to work to have a strong model, but also their weaknesses or limits that will need to be made up for (Chap. 6). This type of question does not exist in monodisciplinary studies, but becomes central to interdisciplinary studies such as the one being proposed here. It is neither possible nor advisable to create a semiotic methodology that is disconnected from the other disciplines that already study animals.

Lastly, I hope this book helped evaluating properly the necessity nowadays of proposing and setting up semiotic studies on animals, from a scientific, societal or ethical point of view.

On one hand, it is true that the issues, especially the ethical ones, linked to animal studies (Chap. 7) have grown exponentially over the past few years because of a combination of a different societal outlook on animals, an awareness of scientific limitations and alarming ecological circumstances. While I do not pretend to bring a universal solution to all of these issues, I believe the introduction of a real semiotic study of animals can help solve certain parts of this multi-faceted crisis that possesses a number of aspects that should be considered grave.

Finally, I hope to have breathed into others the will to build zoosemiotics (Chap. 8) as a discipline in its own right, a bigger discipline that can take in and provide a framework for semiotic methodologies and models applied to animal studies. This discipline is currently growing, it answers needs from both science and society and I can only hope that my work, and the work of all my current and future colleagues will allow us to live up to the expectations that rest upon its shoulders.

References

Alexeev, A., et al. (Eds.). (2017). *Questions Actuelles de la Neurophilosophie* (translation from Russian title). IINTELL: Moscou.

Association Grand Paris Sémiotique. (2019). L'Engagement. *Confrontations sémiotiques*. To be published.

Bertrand, D., & Darras, B. (2019). In Association Grand Paris Sémiotique, *L'Engagement. Confrontations sémiotiques*. To be published.

Bickle, J. (Ed.). (2009). *The Oxford handbook of philosophy and neuroscience*. Oxford: Oxford university press.

Deely, J. (1992). Semiotics and biosemiotics: Are sign-science and life-science coextensive? In T. A. Sebeok, D. J. Umiker-Sebeok, & E. P. Young (Eds.), *Biosemiotics: The semiotic web 1991* (pp. 45–54). Berlin: Mouton de Gruyter.

Delahaye, P. (2017). Cognition, langage, émotion : faut-il sortir du laboratoire ? In A. Alexeev et al. (Eds.), *Questions Actuelles de la Neurophilosophie* (translation from Russian title). IINTELL: Moscou.

Delahaye, P. (2018a). The 38th Albi-Moissac colloquium of French semioticians: Living beings and their environment. *Sign Systems Studies, 46*(2/3), 398–400.

Delahaye, P. (2018b). Founding of the French zoosemiotics society. *Sign Systems Studies, 46*(2/3), 401–402.

Favareau, D., Cobley, P., & Kull, K. (Eds.). (2012). *A more developed sign: Interpreting the work of Jesper Hoffmeyer*. Tartu University Press: Tartu.

Guillaume, A. (2013). Transférabilité du sens d'hier et d'aujourd'hui. *Des mots, des signes, des cultures*. Nanterre, Université de Paris Ouest La Défense.

Guillaume, A. (2014). L'interthéoricité : sémiotique de la transférogenèse. Plasticité, élasticité, hybridité des théories. *Revue PLASTIR, Plasticités, Sciences et Arts, 37*, 1–36.

Hoffmeyer, J. (2008). *Biosemiotics. An examination into the signs of life and the life of signs*. Scranton: University of Scranton Press.

Kull, K. (2012). Advancements in biosemiotics: Where we are now in discovering the basic mechanisms of meaning-making. In S. Rattasepp & T. Bennett (Eds.), *Gatherings in biosemiotics* (p. 18). Tartu: University of Tartu Press.

Machamer, P. (2009). Learning, neuroscience, and the return of behaviorism. In J. Bickle (Ed.), *The Oxford handbook of philosophy and neuroscience* (p. 175). Oxford: Oxford University Press.

Mäekivi, N. (2018). *The zoological garden as a hybrid environment – A (zoo)semiotic analysis*. Tartu: University of Tartu Press.

Rattasepp, S. (2018). *The human Mirror. A critique of the philosophical discourse on animals from the position of multispecies semiotics*. Tartu: University of Tartu Press.

Rattasepp, S., & Bennett, T. (Eds.). (2012). *Gatherings in biosemiotics*. Tartu: University of Tartu Press.

Rothschild, F. S. (2000). *Creation and evolution: A biosemiotic approach*. Edison: Transaction Publishers.

Sebeok, T. A. (1968). *Animal communication: Techniques of study and results of research*. Bloomington: Indiana University Press.

Sebeok, T. A. (1973). *Perspectives in zoosemiotics*. The Hague: Mouton.

Sebeok, T. A. (2001). *Signs: An introduction to semiotics*. Toronto: University of Toronto Press.

Sebeok, T., & Umiker-Sebeok, D. J. (Eds.). (1992). *Biosemiotics: The semiotic web 1991*. Berlin: Mouton de Gruyter.

Sperber, D., & Wilson, D. (1989). La pertinence: communication et cognition. *Paris: les Éditions de Minuit*.

von Uexküll, J. (1934). *Mondes animaux et monde humain. Théorie de la signification*. Paris: Denoël.

Index

P. Delahaye, *A Semiotic Methodology for Animal Studies*, Biosemiotics 19,
https://doi.org/10.1007/978-3-030-28813-6

E

F

G

H

I

J

K

L

M

N

P

MIX
Papier aus verantwortungsvollen Quellen
Paper from responsible sources
FSC® C105338

If you have any concerns about our products,
you can contact us on
ProductSafety@springernature.com

In case Publisher is established outside the EU,
the EU authorized representative is:
Springer Nature Customer Service Center GmbH
Europaplatz 3, 69115 Heidelberg, Germany

Printed by Libri Plureos GmbH
in Hamburg, Germany